The Beginner's Guide to Engineering:
Mechanical Engineering

quantum scientific publishing

The Beginner's Guide to Engineering:
Mechanical Engineering

Mark Huber

quantum scientific publishing

The Beginner's Guide to Engineering: Mechanical Engineering

ISBN-13: 978-1493506453
ISBN-10: 1493506455

Published by quantum scientific publishing

Pittsburgh, PA | Copyright © 2013

Cover design by Scott Sheariss

QUANTUM
SCIENTIFIC
PUBLISHING

Unit One

Section 1.1 – Introduction to Engineering   8

Section 1.2 – Units of Measurement   15

Section 1.3 – Vectors   23

Section 1.4 – Force   31

Section 1.5 – Moments   39

Section 1.6 – Two-dimensional Force Systems   45

Section 1.7 – Three-dimensional Force Systems   53

Section 1.8 – Moments and Moment Vectors   59

Section 1.9 – Couples   63

Section 1.10 – Equivalent Systems   67

Section 1.11 – Systems of Algebraic Equations   73

Section 1.12 – Two-Dimensional Equilibrium   79

Section 1.13 – Three-Dimensional Equilibrium   85

Section 1.14 – Trusses   93

Section 1.15 – Applications   97

Unit Two

Section 2.1 – Force Analysis in Joints   104

Section 2.2 – Section Analysis   111

Section 2.3 – Frames and Machines   117

Section 2.4 – Centroids of Areas   127

Section 2.5 – Centroids of Composite Areas   133

Section 2.6 – Distributed Loads   139

Section 2.7 – Centroids of Volumes and Lines   143

Section 2.8 – Pappus-Guldinus Theorems   149

Section 2.9 – Center of Mass, Simple Objects   153

Section 2.10 – Center of Mass of Composites   159

Section 2.11 – Moment of Inertia   165

Section 2.12 – Moments of Simple Objects   169

Section 2.13 – Rotated and Principal Axes   173

Section 2.14 – Parallel Axis Theorem   177

Section 2.15 – Analysis Application   181

Table of Contents

Unit Three

Section 3.1 – Dry Friction   190

Section 3.2 – Structural Support Beams   199

Section 3.3 – Force and Moment   203

Section 3.4 – Shear Force and Bending Moment   207

Section 3.5 – Distributed Load Moment   215

Section 3.6 – Cables   225

Section 3.7 – Loads Along Straight Lines   229

Section 3.8 – Loads along Cables   235

Section 3.9 – Discrete Loads on a Cable   241

Section 3.10 – Characteristics of Liquids and Gases   249

Section 3.11 – Pressure, Center of Pressure   259

Section 3.12 – Pressure in a Stationary Liquid   263

Section 3.13 – Virtual Work   267

Section 3.14 – Potential Energy   273

Section 3.15 – Applications   281

Appendix

Unit One Answer Key   290

Unit Two Answer Key   302

Unit Three Answer Key   315

Unit One

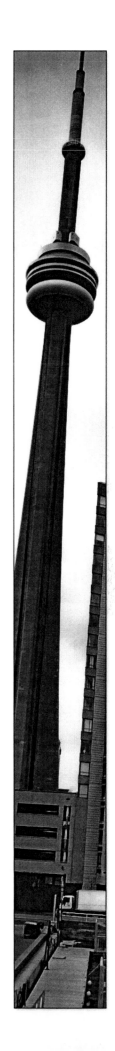

Section 1.1 – Introduction to Engineering   8

Section 1.2 – Units of Measurement   15

Section 1.3 – Vectors   23

Section 1.4 – Force   31

Section 1.5 – Moments   39

Section 1.6 – Two-dimensional Force Systems   45

Section 1.7 – Three-dimensional Force Systems   53

Section 1.8 – Moments and Moment Vectors   59

Section 1.9 – Couples   63

Section 1.10 – Equivalent Systems   67

Section 1.11 – Systems of Algebraic Equations   73

Section 1.12 – Two-Dimensional Equilibrium   79

Section 1.13 – Three-Dimensional Equilibrium   85

Section 1.14 – Trusses   93

Section 1.15 – Applications   97

Section 1.1 – Introduction to Engineering

Section Objective

- Explain the difference between engineering and science

Introduction

This section discusses science and engineering, two occupations which both require creativity, ingenuity and inventiveness. Science can be pursued by anyone at any age. This is how children learn and how big and small questions are initially answered. Once the questions are answered, another type of dedicated creativity is required to figure out how to use this new knowledge. This is engineering.

Engineering versus Science

Science and engineering are related, but different disciplines. They exist on different parts of the continuum that connects basic discovery research and the practical application of the knowledge gained from the basic research. Science generates new knowledge. Scientists do this by asking questions, developing hypotheses and experiments, conducting the experiments and then analyzing the data and communicating the results. Engineering is the practical application of this knowledge to solving problems in the real world. Engineers use the knowledge generated by basic research to do things like build stronger bridges, develop safer chemical compounds, and design faster computer processors.

Science by Definition:

The investigation of natural phenomena through observation, theoretical explanation, and experimentation, or the knowledge produced by such investigation. Science makes use of the **scientific method,** which includes the careful observation of natural phenomena, the formulation of a hypothesis, the conducting of one or more experiments to test the hypothesis, and the drawing of a conclusion that confirms or modifies the hypothesis.

The American Heritage® Science Dictionary Copyright © 2005 by Houghton Mifflin Company. Published by Houghton Mifflin Company. All rights reserved.

Engineering by Definition:

1: the activities or function of an engineer **a:** the application of science and mathematics by which the properties of matter and the sources of energy in nature are made useful to people **b:** the design and manufacture of complex products <software *engineering*> **3:** calculated manipulation or direction (as of behavior) <social *engineering*>

engineering. (2009). In Merriam-Webster Online Dictionary.
From http://www.merriam-webster.com/dictionary/engineering

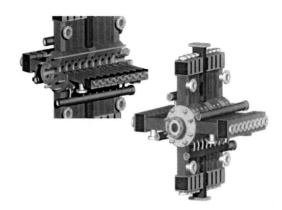

Computer rendering of mechanical designs for the Advanced Photon Source
Image: courtesy of Argonne National Laboratory

Science is an investigation or analytical approach focused on development new understanding of natural phenomena. Science evolved from philosophical studies. Early philosophers tried to explain the things they saw in the world. Philosophers began to develop new ways of asking and answering questions, which led to the development of science and the scientific method. Engineering is a profession that applies scientific knowledge to benefit the world.

Engineering clearly benefits from new scientific developments. Likewise, science also benefits from new engineering developments.

Scientists discovered electricity. The engineers put this to work and created a stable source of transportable energy. This allowed scientists to do interesting experiments using the more complex instruments, which require a stable energy source.

Scientists may have developed the concepts of ink and printing but engineering developed it into a disciplined form allowing for the printing of vast amounts of books.

Physicists and chemists developed the theories behind the semiconductors and computer chips. Engineers applied these theories to develop methods to mass-produce computer chip. They also used their knowledge to optimize the size and shape of the structures, resulting in reduction of the scale of equipment. In 2009, a cellular phone has more computing power than the first desktop computers introduced on the 1980s.

History of Science

Let's look at the history of science briefly. Investigation of the natural world has occurred throughout history. Some records of people observing natural phenomena date back to 3500 BCE. The concept that fire, earth, water, air and idea were the only fundamental elements is from before the time of Socrates (470-366 BCE). This idea about the fundamental elements was considered the one and only truth through the Middle Ages and into the Renaissance (14th through 17th centuries). This theory strongly influenced early European thought. Hinduism, Buddhism and the Japanese used the same elements only replacing "idea" with "Ether" to explain the unseen forces. The Chinese had fire, earth, water, wood and metal.

Distillation
The process of purifying a liquid by first vaporizing it, then condensing it and collecting the resulting liquid.

Crystallization
The process of forming crystals, often by freezing.

Sublimation
The process of changing from solid to gas without going through a liquid phase.

Robert Boyle

Robert Boyle is regarded as the first modern chemist. One of his most well known achievements is a gas law called Boyle's Law, which describes the proportional relationship between the absolute pressure of a gas and its volume in a closed system with a constant temperature.

Science builds upon knowledge generated by earlier scientists. The knowledge generated throughout history has allowed us to develop the sciences and technologies that make our current way of life possible. Let's look at a few of these now.

One of the earliest sciences was alchemy. Alchemy is often regarded as the origin of modern chemistry. Alchemy began not only as a study of nature, but also as a philosophical and spiritual practice. Alchemy began more than 2,500 years ago and is still practiced in some cultures.

The Alchemist Lab

In their investigations of nature, early alchemists focused on improving their practical skills of transmutation (the production of gold from base metals), They also searched for the elixir of life and the source of immortality.

Arabic cultures, in medieval times, produced systematic surveys of alchemy by blending their own pharmacological (herb/drug-based treatment) tradition with Greek, Indian, and Chinese work. They also improved techniques for **distillation**, **crystallization** and **sublimation**. The term "alchemy" is believed to be from Arabic. At this same time, in medieval Europe, alchemy was considered a heretical (aberrant and unacceptable religious) practice and advances came from isolated efforts by people working alone.

In the 1500s, the concepts of physics, through astronomy studies, became recognized as a separate discipline from the other natural sciences. In the 1600s, chemistry became more recognized as a scientific pursuit as Robert Boyle published his text describing chemistry apart from alchemy.

Starting in the 1700s and 1800s, as more discipline was applied to the various branches of science, primarily chemistry and physics, the field of science began to diverge from the field of philosophy. This occurred as the scientific method took hold and more discipline was applied to the study. By the early 1900s, the concept of science as we know it now was established.

Now let's look at engineering. Examples of applications of the concepts of civil and mechanical engineering can be found as early 200 BCE in ancient Greece. Examples include the Roman aqueducts, the Parthenon, the lost city in India, and the principles (buoyancy, for example) that were defined by Archimedes. In the 1200s, the Arabic cultures extensively developed the concepts of mechanical devices and tools. However, it was not until the 1800s that mechanical engineering separated from the general engineering disciplines and titles like "engineer" were used to define a professional class.

Bonnie Dunbar (born 1949) is a mechanical engineer and an astronaut.
Image: Courtesy of NASA

The concepts of electrical engineering can be found as early as the 1600s with William Gilbert. These concepts were developed more fully in the early 1800s by Alessandro Volta, George Ohm and Michael Faraday. In the late 1800s, James Maxwell and Heinrich Hertz developed the field of electronics further expanding electrical engineering.

Over time, more areas of science are being applied in the world. This expands the overall field of engineering with new sub-disciplines, including:

- **chemical engineering** – the application of engineering principles to problems involving chemistry, such as drug manufacturing and the development of household cleaning products.

- **civil engineering** – the application of engineering principles to problems of civilization, such as road construction, sewage treatment, and building design.

- **computer engineering** – the application of engineering principles to problems involving computers, such as processor speed, battery life, and memory.

- **aeronautical engineering** – the application of engineering principles to flight, including aircraft design.

Archimedes

Archimedes was an ancient Greek scientist who first described the principles of buoyancy. He also developed some basic physics concepts, designed machines, and was an extraordinary mathematician.

- **aerospace engineering** – the application of engineering principles to space flight and travel. Examples include the space shuttles, satellites and the International Space Station.

- **genetic engineering** – the application of engineering principles to questions of genetics. Examples include genetically modified foods that are designed to be resistant to pests or disease.

- **software engineering** – the application of engineering principles to the development of new software, including maximizing the efficiency of the code so the computer processors are able to work as quickly as possible.

- **solar engineering** – the application of engineering principles to collecting, storing and using energy collected from the sun to heat homes and do other work.

Branches of the Natural Sciences

The natural sciences generate knowledge that is used by engineers to solve practical problems. The natural sciences are those that study the natural world and include disciplines ranging from astronomy to physics. The knowledge generated by the natural sciences is used by engineers to solve practical problems that will benefit humans. The result may be a new type of plant, a new airplane design, or a car that runs on air. We don't know what will be developed until someone has translated the basic knowledge into a practical application that solves some problem.

Branch of Natural Science	Description
Astronomy	Astronomy is the study of the stars and planets and the interactions between these bodies.
Biology	Biology is the study of living things, plants, animals and humans.
Chemistry	Chemistry is the study of the fundamental makeup of the elements on the planet and their interactions.
Ecology	Ecology is the study of the ecosystem, the impact of man on the environment and searches for solutions to the challenges of pollution and waste on the earth.
Geology	Geology is the study of the macro scale physical aspects of the planet, the minerals, the continents, and plate tectonics.
Physics	Physics is the study of nature with the goal being to explain the reasons that things behave the way they do. Along with Newtonian mechanics and quantum mechanics, physics also covers thermodynamics, electromagnetism and the new theories of relativity.

Branches of Engineering

Engineering takes the knowledge generated by the natural sciences and uses engineering principles to solve practical problems. Engineers apply the knowledge generated by the natural sciences, which is why engineering is considered an applied science. It can be very difficult to translate a finding made in a laboratory to a scale where a product can be consistently produced or a principle applied on a large scale. Engineers often develop the processes required to translate these basic processes so they can be used to produce a product.

Branch of Engineering	Description
Aeronautical and Aerospace Engineering	This discipline deals with the design and construction of airborne craft and vehicles that will be launched into space.
Chemical Engineering	This discipline primarily focuses on using chemicals in industrial applications.
Civil Engineering	This focuses on the structures needed for modern society, which includes buildings, bridges, roads, highways and large infrastructure projects like airports and hydroelectric facilities.
Computer Hardware and Software Engineering	This discipline applies the concepts of mathematics to computation through hardware (computer hardware engineering) and computer software (software engineering).
Electrical engineering	Electrical Engineering applies the principles of physics to industrial uses. This includes magnetics, electronic circuits and Radio Frequency (RF) communication.
Genetic Engineering	Genetic engineering uses the principles of genetics and biology in combination with various engineering disciplines. The engineering disciplines include electrical, mechanical and hydrodynamic engineering.
Mechanical Engineering	This branch of engineering applies the principles of mechanics, which is a branch of physical science dealing with mechanical interactions between bodies and the interactions of forces and motion. Statics is a branch of mechanics focused on the analysis of the various forces active in a rigid structure when it is in equilibrium. Dynamics is a branch of mechanics focused on the analysis of the interrelation between forces and motion of bodies.

Other aspects of engineering that fall under mechanical engineering are energy production, fluid mechanics, kinematics, general mechanics, heat and mass transfer and thermodynamics. Applications include the design and analysis of vehicles, structures, manufacturing processes, medical instrumentation, robotics, motors, generators, energy sources, etc.

Concept Reinforcement

1. Describe the basic concept of science

2. Describe the basic concept of engineering

3. Explain the difference between science and engineering

4. Explain what is studied in each of the following fields of science:

   biology, ecology, and physics

5. Explain the focus of each of the following fields of engineering:

   mechanical engineering, civil engineering, and genetic engineering

Section 1.2 – Units of Measurement

Section Objective

- Describe the systems of units and unit conversions

Introduction

This section presents the basic units in the two most prominent systems of measures, the International System (SI) of units and the English Customary system of units. Many countries have adapted the SI system because of the simple advantages related to the scaling of the sizes being a factor of 10 difference. Additionally, the SI system does not require converting between sets of units, such as gallons and pints, which reduces the potential for errors.

Why are systems of measurement important?

Engineering requires both precision and accuracy. Many applications of engineering have very low tolerance for error, often in the hundredths of millimeters for machined metal pieces. This concept also applies to circuits, computer chips, fluid measures, temperature, time, and any other measured variable. Two main systems of measurement are used in the US: The Metric System and the US Customary System. The metric system uses SI Measurement units and is the measurement system accepted for scientific and engineering use. The US Customary System is a system of measurement that came from Europe when the first settlers came to the country. These units of measure are not based on any particular starting unit. A foot was actually the measure of a person's foot, so people with different size feet would measure different actual distances but count the same number of feet. You can begin to see where the metric system might be easier to use. Every starting measure can be multiplied or divided by ten to get to the next larger or smaller level of measurement.

Hy Tran examines a kilogram sample in a mass comparator
at Sandia's Primary Standards Laboratory. Image courtesy of Sandia Laboratories.

The SI (International System) of units contains seven base units.

The SI base units have been defined as the standard by which all measurements can be compared. The SI base units, along with the entire system, were developed in 1960 to provide a uniform system of measures. Before that time, two systems had developed: the meter-kilogram-second system and the centimeter-gram-second. These were two systems that developed to reduce the variation in measures. The SI developed to further reduce the variation in measures internationally.

Physical Quantity	Unit	Symbol
Length	meter	m
Time	second	s
Mass	kilogram	kg
Temperature	Kelvin	K
Amount of Substance	mole	mol
Electric Current	ampere	A
Luminous intensity	candela	cd

Derived Units Physical Quantity	Unit	Symbol
Area	square meter	m^2
Dry volume	cubic meter	m^3
Liquid volume	liter	L

The Convention du Mètre of May 20, 1875 is an international treaty that established three organizations to oversee the keeping of metric standards. These organizations focus on accurately defining these various measures and finding more accurate methods to independently describe these base units. For instance, these organizations are:

- The General Conference on Weights and Measures, which meets every four to six years with delegates from all member states.

- The International Bureau of Weights and Measures, which is an international metrology (study of measurement) center located in France.

- The International committee for Weights and Measure, which is an administrative committee that meets yearly at the International Bureau of Weights and Measures.

The US Customary System and Measurement Units

The U.S. Customary System of units is also called the English, Imperial or standard system of units. There are several common units for some of the physical quantities.

The English system of units has existed for centuries. This system began in the Roman Empire and may even pre-date that.

Because of this extensive and varied history, the units in the English System have different and interesting roots. Records show that the term "inch" in one society was the term for the width of a thumb, but in a different society was the term for the length of 3 barley corns end to end. The term "foot" originally meant the length of a person's foot and the term "mile" was the distance of 1,000 paces of a Roman Legion. Here a "pace" is the length of two steps (right and left).

Physical Quantity	Unit	Symbol
Length	inch	in
	foot	ft
	yard	yd
	mile	mi
Time	second	s
Mass	Slug	Slug
Temperature	Farenheit	°F
Amount of Substance	mole	mol
	pound-mole	lb-mol
Electric Current	ampere	A
Luminous intensity	Candlepower	cd
Dry volume	cubic inch	cu in
	cubic foot	cu ft
	cubic yard	cu yd
Liquid volume	pint	pt
	quart	qt
	gallon	gal

Over time, the terms used for different physical quantities were adapted so they were convertible with each other.

Unit Conversions

Length

To convert between the SI units (meters) and the US Customary units:

> 1 meter (m) = 39.37 in (inches) = 1.0936 yd (yard)
> 1 foot (ft) = 0.3048 m (meters)

From the meter, with the use of scientific notation, smaller divisions are defined such as centimeter (cm) being 1/100th of a meter or micrometer (μm) being 1 millionth of a meter.

To convert between centimeters and inches:

1 centimeter (cm) = 0.39370 in (inches)
1 inch = 2.54 cm (centimeters)

Larger divisions are periodically used, as well, with one kilometer (km) equal to 1,000 meters, often used.

To convert between kilometers and miles:

1 kilometer (km) = 0.622 mi (miles)
1 mile (mi) = 1,609 m (meters) = 1.609 kilometers (km)

Time

Time in seconds is universal, with minutes (60 seconds), hours (60 minutes = 3600 seconds) and days (24 hours = 1440 minutes = 86400 seconds).

Mass

The SI system mass unit is the kilogram (kg). Note that mass is not weight. While mass is constant if you are on the Earth or the Moon, weight is different because it is the influence of the gravitational pull towards the Earth (or the Moon) on the mass of an object that defines weight.

The kilogram (kg), in combination with scientific notation, has smaller mass quantities. While a kilogram = 1,000 grams, measures can be in a single gram (g), microgram (μg) or smaller or larger.

The conversion between SI mass units and US customary mass units is rare. A conversion between pounds (force) and kilograms (mass) is common, with 0.4536 kg per pound.

Temperature

Kelvin is a measure of temperature that is related to the Celsius scale. Kelvin has been defined as a scale that begins at a temperature that is the theoretical point at which there is no movement of matter. This is called absolute zero and has been established at -273.15 °C.

In the Celsius scale, water freezes at 0 °C and boils at 100 °C, a 100 °C difference.

In the Kelvin scale, water freezes at 273.15 K and boils at 3273.15 K, a 100 K difference between the freezing point and boiling point.

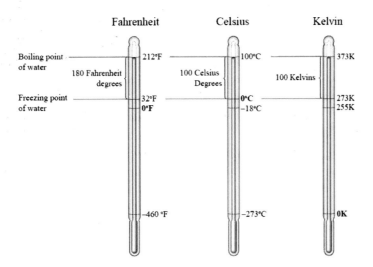

Note that the "degree" symbol (°) is not used when describing temperature in terms of Kelvin.

Amount of substance

The mole equals 6.0221367×10^{23}, which is usually shortened to 6.02×10^{23}. The mole is a number, similar to the term dozen, which represents a quantity of anything. In the way we can say a dozen eggs, chemists say a mole of a substance, such as a mole of hydrogen gas.

Note that this is a very large amount of a substance. However, when chemists talk about atoms, molecules and compounds, these are very small individual units. Therefore, even though a mole is a large amount of the substance, when we talk about a mole of hydrogen, its mass is just 1 gram (1.008 g/mol).

Electric current

Electric current is measured using the ampere (A), an amount of electric charge per second. It is the flow (movement) of electric charge.

One ampere is one coulomb of electric charge per second flowing through an imaginary 2-dimensional plane (a cross section) of a conductor. 1 coulomb is approximately 6.242×10^{18} electrons.

Note that the actual path of electrons through a conductor is quite random. Even though it sounds like the electrons are moving in a single direction like water in a stream, a better metaphor is that electrons move like a 3-dimensional batch of marbles bouncing down a slightly sloping ramp with some of the marbles moving in every direction.

Luminous Intensity

Luminous intensity is a measure of the power emitted by a light source in a particular direction. This measure is based on a standardized model of the sensitivity of the human eye.

A common candle emits approximately 1 candela (1 cd) while a 100 watt incandescent bulb emits approximately 120 candela (120 cd).

1 candlepower (US units) = 1 candela (metric units)

Metric Prefixes and Scientific Notation

In order to more readily express a full range of measures when describing things, the metric system uses prefixes as shortcuts to scientific notation that express a linear relation to the powers of 10.

From the following table these very small or very large numbers can be rewritten in several much more convenient formats. Most people have heard of the terms one thousand (1,000), or one million (1,000,000) or even one billion (1,000,000,000).

These can be rewritten using various prefixes or notations.

One thousand (1,000) can also be written as one kilo or one "k." When you say one kilo or one k, you mean that you have one chunk of something the size of 1,000 individual units. For example: one kilometer (or 1 km) means 1,000 meters. This is similar for one million (1,000,000), which can be rewritten as one mega or one "M" or one billion (1,000,000,000), which can be rewritten as one giga or one "G."

One hundredth of a unit can also be written as one centi. If you have one centi or one c of something, this means that you have one chunk of something 1/100th the size of the individual units. For example: one centimeter (or 1 cm) means 1/100th of a meter. This is similar for micro (μ), like micrometer or μm, which can be rewritten as 1/1,000,000th of a meter or nano (n), like nanometer or nm, which can be rewritten as 1/1,000,000000th of a meter.

Examples of this notation:

> 50 μm is written as fifty micrometers
> 3nm is written as three nanometers
> 10cm is written as ten centimeters
> 100pg is written as one hundred picograms
> 25 μs is written as twenty five microsecond

Prefix	Symbol	Magnitude	Meaning (multiply by)
tera-	T	10^{12}	1,000,000,000,000
giga-	G	10^{9}	1,000,000,000
mega-	M	10^{6}	1,000,000
kilo-	k	10^{3}	1,000
hecto	h	10^{2}	100
deka	da	10	10
-	-	-	-
deci-	d	10^{-1}	0.1
centi-	c	10^{-2}	0.01
milli-	m	10^{-3}	0.001
micro-	μ (mu)	10^{-6}	0.000001
nano-	n	10^{-9}	0.000000001
pico	p	10^{-12}	0.000000000001
femto-	f	10^{-15}	0.000000000000001

There is another convenient way to express large and small numbers. This is using the power of ten. When we have one or ten bananas it is easy to write. These quantities can be rewritten in a different format using the power of ten.

> One is ten to the power zero (10^{0})
> Ten is ten to the power one (10^{1})
> 100 (2 zeros after the digit 1) is ten to the power two (10^{2})
> 1,000 (3 zeros after the digit 1) is ten to the power three (10^{3}).

We can also use this to express small numbers. Recall that a negative power is essentially saying this is a number in the denominator.

> 10^{-3} is one divided by one thousand (3 zeros after the digit 1). This is the same as 0.001 (the decimal point is moved 3 points to the right after the 1)
> 10^{-9} is one divided by one billion (9 zeros after the digit 1). This is the same as 0.000000001 (the decimal point is moved 9 points to the right after the 1).

Examples of this notation:

> 50 μm is written as fifty micrometers or as 50×10^{-6} m
> 3nm is written as three nanometers or as 3×10^{-9} m
> 10cm is written as ten centimeters or as 10×10^{-2} m
> 100pg is written as one hundred picograms or as 100×10^{-12} g
> 25 μs is written as twenty five microseconds or as 25×10^{-6} s

Concept Reinforcement

1. Explain the reasons for standard units of measure to exist.

2. List the seven base units of the SI system.

3. Convert the measure of 21 feet to meters, kilometers, centimeters and micrometers using scientific notation first with symbols and then using powers of 10.

4. Convert the measure of 300 feet to meters, kilometers, centimeters and micrometers using scientific notation first with symbols and then using powers of 10.

Section 1.3 – Vectors

Section Objective

- Explain vectors and describe how they are used

Introduction

This section presents the concept of scalars and vectors as types of expressions of measurement. A scalar expresses only a magnitude (amount). Examples of scalar expressions are "my cup is full" or "my battery is weak." Direction is not relevant in either case. The direction of my cup's volume of liquid is of no value and the direction of the strength of the battery is also of no value.

A complete expression of a vector includes both a magnitude and a direction. Examples of vector expressions are "I dropped my cell phone" or "the birds flew north," In both these cases the direction is important information. In the first expression, the direction is implied (dropping = downward). However, note there is no magnitude in either of these expressions. Therefore, they are not complete vectors. Complete vectors are "my cell phone dropped at 9.8 m/s" and "the birds flew north at 15 mph."

Scalars and Vectors

All measurements express a magnitude with the magnitude being the amount of the physical quantity. Each of the following examples presents a magnitude as a way to compare these events to other events.

Example 1: Magnitude

She is sick and has a fever of 101°F

He weighs 185 lbs.

The balloon has 15 liters of helium inside.

The trip odometer on the car showed we traveled 105 miles.

No direction specified in any of these examples. Someone's fever can only increase or decrease on the temperature scale. Someone's weight can only increase or decrease on the weight scale, and the amount of helium inside the balloon can only increase or decrease. This is complete information. The direction of the fever, the weight or the volume is not relevant. Only with the car can we suggest that a direction may be relevant.

> **Magnitude**
>
> Amount. Magnitude is a scalar quantity, which means it measures only the amount of something, not its direction.

Comparison of one unit to one thousand units

It is possible to describe the direction the car traveled, for instance, westward. In this case, adding the direction adds relevant information. It helps to know that the car traveled to the west because we will know where it has come from and where it has gone.

Example 2: Magnitude and Direction

A group starts out in their car from Austin, Texas and they travel 105 miles. They stop at a rest station and ask how much further to San Antonio. At this point, we need to ask where they are. It will help a great deal to know the direction of their travel from Austin. If they traveled straight east, the distance to San Antonio, which is west of Austin, has increased. If they traveled straight west, they may have passed San Antonio, but they will be closer than they would be if they traveled east.

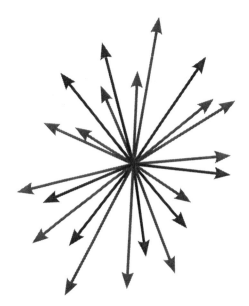

Image showing vectors - each arrow indicates magnitude (length) and direction (angle).

When the measurement only provides a magnitude it is called a scalar quantity. The fever, weight and volume are scalar quantities, or scalars.

Vector

Amount + Direction. A vector provides information on both the amount of something (scale) and its direction.

When the measurement provides a magnitude and a direction that is relevant it is called a vector quantity. The length of travel and the direction are both relevant and this combination is called a vector quantity or just vector.

Algebra Operations on Scalars

When we add or subtract scalars, this is adding or subtracting a magnitude. This is what is normally done in algebra.

- We can add the temperature of two objects: 101°F + 51°F = 152°F

- We can add the weights of two people: 185 lb–115 lb = 70 lb

- We can add the volume of two spaces: 15 L + 5 L = 20 L.

These answers are correct and this is all the relevant information available. The direction is not relevant information. This is the same for multiplication and division.

With vectors, these simple functions take on a different meaning.

Adding Vectors

When adding vectors, the direction is relevant information and affects the results of the calculation.

Example3: Adding Vectors

Vector 1 (V1) has a magnitude 10 with a direction of east.
Vector 2 (V2) has a magnitude of 10 with a direction of east.
Calculate the resultant vector.

Solution:

We draw the first vector. Then, starting where the first vector ends, we draw the second vector. These two vectors, because they are in the same direction, are simply added together. Vector 1 is 10 units to the east and vector 2 is 10 additional units to the east. The magnitude of the resultant vector is 10 + 10 = 20 and the direction is still east. The answer is written as 20E.

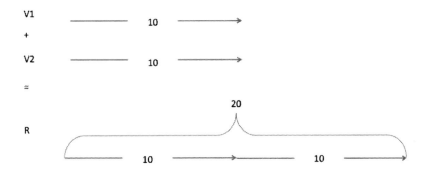

Example 4: Adding Vectors

Vector 1 (V1) has a magnitude of 10 with a direction of east.
Vector 2 (V2) has a magnitude of 10 with a direction of west.
Calculate the resultant vector.

Solution:

We draw the first vector. Then, starting where the first vector ends, we draw the second vector. In this case, the second vector retraces the path of the first vector – back to the origin. Because they are in opposite directions, the second vector is subtracted from the first. Vector 1 is 10 units to the east and vector 2 is 10 units to the west. We can solve this in two ways, which both depend on the direction.

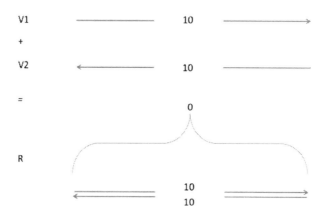

First, there is a connection between the direction and the magnitude.

We know that east and west are opposite directions. If we travel a positive 10 steps to the east, this is equivalent to a negative 10 steps to the west. Both of these descriptions are correct. This assumes that travel to the east is in the positive direction, so the positive steps are going towards the east.

Therefore, Vector 1 is a positive 10 steps to the east and Vector 2 is a negative 10 steps to the east.

We can also define that travel to the west is in the positive direction, so the positive steps are going to the west. In this case, Vector 1 is a negative 10 steps to the west and Vector 2 is a positive 10 steps to the west.

Either way, Vector 1 is in the opposite direction of Vector 2. Because the magnitudes are the same and the directions are opposite, they cancel each other and the resultant vector is zero.

Example 5: Adding Vectors

Vector 1 has a magnitude of 10 with a direction of east.
Vector 2 has a magnitude of 10 with a direction of north.
Calculate the resultant vector.

Solution:

We draw the first vector. Then, starting where the first vector ends, we draw the second vector. In this case, the second vector is in a direction that is 90° from the first vector.

This requires the use of trigonometry and the Pythagorean Theorem for right triangles, where $A^2 + B^2 = C^2$. In this case, Vector 1 is A, Vector 2 is B and the resultant vector is C. Solve for C (the magnitude of the resultant vector):

$$C = \sqrt{A^2 + B^2} = \sqrt{10^2 + 10^2} = \sqrt{200} = 10\sqrt{2} = 14.14$$

Because they have the same magnitude and their directions are 90° apart, the result is a 45° angle towards the northeast.

The angle can be determined using various trigonometry functions:

Tangent: Tan (10/10) = 45°

Sine: Sin (10/14.14) = 45°

Cosine: Cos (10/14.14) = 45°

The resultant vector is 14.14 ∡ 45° NE.

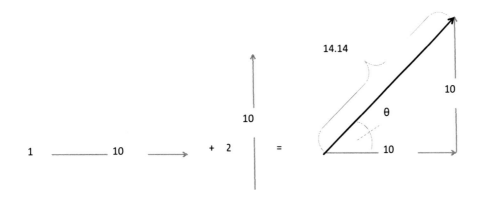

Now we'll expand on the concept of positive and negative direction for the vectors. There is a positive-negative scale for the east –west direction. There is a positive-negative scale for the north-south direction.

Each vector, including the resultant vector, can be stated using the various orientations of the axes.

Some of the possible correct solutions are:

- Vector 1 is a positive 10 steps to the east and vector 2 is a positive 10 steps to the north.

- Vector 1 is a negative 10 steps to the west and vector 2 is a positive 10 steps to the north.

- Vector 1 is a positive 10 steps to the east and vector 2 is a negative 10 steps to the south.

- Vector 1 is a negative 10 steps to the west and vector 2 is a negative 10 steps to the south.

Note that the origin (0° angular direction) for the vector angle is directed horizontally, in this case, straight east.

Subtracting Vectors

As can be seen from the examples in the Adding Vectors section, the operation depends upon how the orientation of the direction is defined. Will "east" be the positive direction? Will north also be defined as a positive direction?

With this in mind, subtraction of vectors is very similar to addition of vectors. In fact, it is best to change all vector subtraction problems into vector addition problems. This is done, as presented above, by reversing the direction of the vector to be subtracted. The direction is then "positive" and the magnitude is negative.

Example 6: Subtracting Vectors

Vector 1 has a magnitude 20 with a direction of east.
Vector 2 has a magnitude of 12 with a direction of east.
Calculate the resultant vector if vector 2 is subtracted from vector 1.

Solution:

We draw the first and second vectors. We change the subtraction into an addition by reversing the direction of the second vector. Now we have a simple addition problem.

Starting at the point of the arrow, where the first vector ends, we draw the second vector. This is in the opposite direction. Together we have 30 steps to the east and 12 steps back to the west. The magnitude of the resultant vector is $20 + (-12) = 8$ and the direction is east.

Example 7: Subtracting Vectors

Vector 1 has a magnitude of 10 with a direction of east.
Vector 2 has a magnitude of 10 with a direction of north.
Calculate the resultant vector if we subtract vector 2 from vector 1.

Solution:

We draw the vectors. Then we create a vector addition by reversing the direction of Vector 2. Then we draw the second vector starting where the first vector ends. In this case, the second vector is in a direction that is 90° from the first vector.

This requires the use of trigonometry and the Pythagorean Theorem for right triangles, where $A^2 + B^2 = C^2$. In this case, Vector 1 is A, Vector 2 is B and the resultant vector is C. Solve for C (the magnitude of the resultant vector):

$$C = \sqrt{A^2 + B^2} = \sqrt{10^2 + 10^2} = \sqrt{200} = 10\sqrt{2} = 14.14$$

Because they have the same magnitude and their directions are 90° apart this is a 45° angle towards the southeast.

The angle can be determined using various trigonometry functions:

Tangent: Tan (10/10) = 45°

Sine: Sin (10/14.14) = 45°

Cosine: Cos (10/14.14) = 45°

The resultant vector is 14.14 ∡ 45° SE.

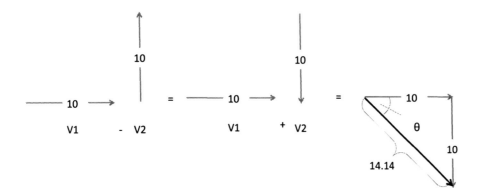

Concept Reinforcement

1. State the difference between a scalar quantity and a vector quantity.

2. Add two vectors; vector 1 is 15 units west, vector2 is 5 units west.

3. Subtract vector 2 (2 units south) from vector 1 (10 units north)

4. Add vector 1 (5 units north) to vector 2 (10 units west).

Section 1.4 – Force

Section Objective

- Discuss forces and free-body diagrams

Introduction

This section presents the concept of a force and describes the general types of forces that an object can experience. Contact forces, action-at-a-distance forces and the governing laws of motion first presented by Isaac Newton are also presented. A graphical representation of forces acting on an object is also discussed. This is called a free-body diagram where the forces are presented as arrows with the arrow length equal to the relative size of the force and the arrow direction showing the direction of the force. The free-body diagram is a useful tool for analysis of force and object interactions because it presents everything in a concise visual form and because the forces are presented as acting from the center of the object, which simplifies the calculations.

A Definition of Force

Force is the term that describes the effort to push or pull an object, or the amount of effort required to move an object over a distance. Something gets pushed or pulled and it moves. If it was originally at rest and it moved due to the push/pull then there is a change in velocity.

If we compare the change in velocity (Δv) over the same time interval (Δt), for two objects:

- Object 1. Δv is from zero to some final velocity v_{f1}

- Object 2. Δv is from zero to some final velocity v_{f2}; where $v_{f2} > v_{f1}$.

Then we can say that, since the change in velocity (Δv) is larger for object 2 (over the same time interval), this inplies that more force was applied to make this happen. Force is proportional to the change of velocity which is more formally called acceleration and therefore force is proportional to acceleration.

Because both the amount and the direction of this push or pull carry relevant information, this inplies that the force is directional, which means force is a vector quantity, with magnitude (amount) and direction. The amount and direction are both relevant aspects of the acceleration and therefore acceleration is also a vector.

One more relevant aspect when talking about forces is the physical size of the object. If we find an object 3 that has more mass than object 2 and subject it to the same push that was delivered to object 2 we know that object 3 will not have as much change of velocity. Therefore force, acceleration and mass are linked.

We can say that force is proportional to both mass and acceleration.

The formula for force is: Force = mass × acceleration.

Force: $F = ma$

Where:
F = force in newtons (N)
m = mass in kilograms (kg)
a = acceleration in meters/second squared (m/s^2)

In the SI system:

- The unit of mass (m) is the kilogram (kg)

- The unit of acceleration (a) is m/s^2 (meter per second squared)

- The unit of force (F) is the newton (N) where

- 1 newton = 1 kg × m/s^2 = 1 kilogram × meter/second^2

In the English system:

- The unit of mass (m) is the slug (slug)

- The unit of acceleration (a) is in ft/s^2 (feet/second^2)

- The unit of force (F) is the pound (lb) where

- 1 pound = 1 slug × ft/s^2 = 1 slug × feet/second^2

How can we tell if a force is present? In the relation presented in the formula $F = ma$, note that a force cannot exist without an acceleration or a mass.

Therefore, the object would need to have mass and it would need to have an acceleration (a change in its velocity) for force to be acting on it.

Note also that since acceleration is a vector and mass is a scalar (no direction defined), this equation is effectively a vector multiplication that results in force being a vector that is a multiple of the acceleration vector. The direction of the acceleration vector dictates the direction of the force vector while the magnitude of the mass and the acceleration both influence the magnitude of the force.

Therefore if acceleration is acting on an object that has mass there is a force with magnitude of $F = ma$ and a direction that is the same as the acceleration (a).

Categories of Force

Forces are usually broken into two groups: **contact** forces and **action-at-a-distance** forces. **Contact** forces are involved when two objects interact and physically contact each other. **Action-at-a-distance** forces exist everywhere and exert a push or pull on objects without having a physical contact.

Contact Forces	Action-at-a-distance Forces
Frictional Force	Gravitational Force
Tension Force	Electrical Force
Normal Force	Magnetic Force
Air Resistance Force	
Applied Force	
Spring Force	

A Definition of Net Force

If acceleration is acting on an object in several different directions, then it is possible to talk about several forces acting on the object. Each force is in the same direction as the associated acceleration. Because force is a vector, we can use vector addition when more than one force is present and determine a "net force," which is the sum effect of all the various forces acting on that object.

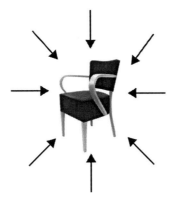

When no forces push or pull on the object or the object is not moving, there is no net force, or more accurately, the net force is zero.

When a force pushes or pulls on an object, the net force that be greater than zero. You need to know two things to calculate net force: the mass of the object and the acceleration that is acting on the object.

If the mass $(m) = 30$ kg and acceleration $(a) = 10$ m/s^2, then force $F = ma = (30\,\text{kg})(10\,\text{m/s}^2) = 300$ kg $\times$ m/s$^2 = 300$ N (newtons)

We can also determine the acceleration by noting the change in velocity, since acceleration $(a) = \Delta v/\Delta t = (v_f - v_i)/t$.

If the initial velocity (v_i) is 15 m/s and the final velocity (v_f) is 55 m/s and this happened in time (t) of 10 s, the acceleration (a) is = (55 m/s–15 m/s)/10 s = 40/10 = 4 m/s^2.

If mass (m) = 30 kg, then force, $F = ma = $ (30 kg)(4 m/s^2) = 120 kg × m/s^2 = 120 N (newtons).

When several forces push or pull on the object, the net force is the vector sum of the disparate forces.

When the two vectors are in the same direction, the problem becomes a simple addition of the magnitudes. If force 1 (F_1) is 30 N and force 2 (F_2) is 35 N and they act in the same direction, then the sum is $F_1 + F_2 = $ 30 N + 35 N = 65 N.

When several forces push or pull on the object in opposite directions, the net force is the vector sum of the disparate forces. These forces can be arranged to be a simple addition of two vectors in the same direction, one with a positive magnitude and the other with a negative magnitude.

If force 1 (F_1) is 30 N to the west and force 2 (F_2) is 35 N to the east, then the sum can be written several ways.

If we use west as the positive direction, then east is the negative direction. F_1 is positive and F_2 is negative.

Since the forces have a magnitude and direction, the forces will then be:

$F_1 = $ $^+$30 N west and $F_2 = $ $^-$35 N west

The sum is: $F_1 + F_2 = $ $^+$30 N + ($^-$35 N) = 5 N to the west or $^+$5 N to the east.

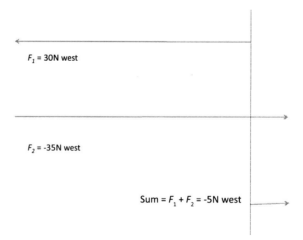

Because a vector with a negative magnitude in the direction west can also be described as a vector with a positive magnitude in the direction east, this can also be stated as $^+$5 N to the east. This assumes that the direction "west" is exactly opposite the direction "east."

If, on the other hand, we use east as the positive direction, then the directions (and signs) change and west is the negative direction. F_1 is negative and F_2 is positive.

Since the forces have a magnitude and direction, the forces will then be:

F_1 = 30 N east and F_2 = $^+$35 N east.

The sum is: $F_1 + F_2$ = 30 N + ($^+$35 N) = $^+$5 N east, which can also, of course, be stated as 5 N to the west.

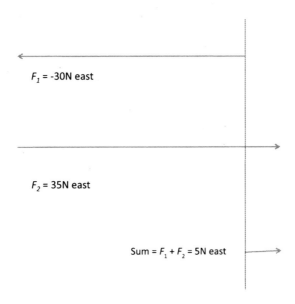

F_1 = -30N east

F_2 = 35N east

Sum = $F_1 + F_2$ = 5N east

The results are exactly the same because these are the same forces, acting the same directions. We just used different frameworks, which are different points of view. One framework assigned "west" as the positive direction and the other framework assigned "east" as the positive direction.

Free-body Diagrams

A free-body diagram is a graphical representation of the various forces acting on an object in a specific situation and at a specific time. It is a vector diagram where all the forces, which are vectors, have their origin at the geometric center of the object.

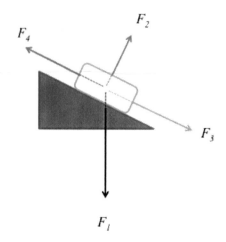

Free body diagram of an object on a ramp

In the illustration, four forces are shown, with each force acting in a different direction.

The order of the number is important here.

- Force F_1 is usually the force of gravity (F_g).

- Force F_2 is usually the normal force (F_N), the force of the ramp against the object, which opposes that fraction of the gravitational force.

- Force F_3 is usually the fraction of the force of gravity ($F_g \sin\theta$) that acts on the object at the angle of inclination of the ramp. Here θ is the angle of inclination.

- Force F_4 is usually the force of friction (F_f), which is related to the normal force and the properties of the material.

In the illustration, the dashed lines at the "tail" of each force are shown only to demonstrate that the forces all originate from the geometric center of the object. Normally, the force vectors are only shown from the edge of the object outward and the magnitude of the force is shown using the length from the edge of the object to the tip of the force arrow.

Free-body diagrams follow several guidelines;

1. The object is drawn as a simple geometric shape, usually a box

2. All forces are shown as arrows

3. The forces are shown acting on the outer edge of the object

4. The origin of each force is the geometric center of the object

5. The length of the arrow shows the relative magnitude of the forces

6. The direction of the arrow shows the direction of the force

7. Each force vector is labeled

An object may experience a single force, two forces, or many forces. Therefore, the free-body diagram of that situation would also show one, two or many forces.

Concept Reinforcement:

1. What is the definition of force?

2. Explain how to determine acceleration.

3. Explain why force is a vector.

4. List the guidelines for creating a free-body diagram.

5. Solve the following problems:
   5a: Mass=10 kg; Acceleration=30 m/s². Solve for force in newtons.

   5b: Mass=52 kg; Acceleration=75 m/s². Solve for force in newtons.

6. Create a free-body diagram
   6a. An object is on a flat plane. It has a force of gravity = 980 N, a normal force which opposes the force of gravity and a force from the left of 100 N.

   6b. An object is on a ramp with an incline. It has a force of gravity = 1,000 N, there is a normal force which opposes the fraction of the force of gravity, a force down the ramp which is a fraction of the force of gravity and a force of friction of 100 N which is acting opposite the direction of movement. Define the magnitude of the forces as much as possible.

   6c. Create a free-body diagram of a person on a snowboard sliding down a hill.

Section 1.5 – Moments

Section Objective

- Explain moments and their uses

Introduction

This section presents the concepts of a mechanical moment, which is a more generalized term for torque. A moment (and a torque) is the effect caused by a force acting at a radial distance from a point. This force acts perpendicular to the radial line connecting the point and the force. The moment is a vector, with a direction that is always perpendicular to the plane described by the force and the radial line. Oftentimes, the direction of the moment can be simplified to "clockwise" or "counterclockwise," but this terminology quickly loses its relevance when an object experiences moments in more than two dimensions.

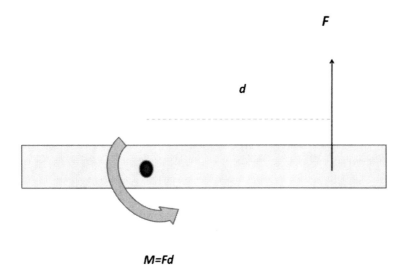

$$M=Fd$$

A Definition of Moment

"Moment," in mechanical engineering, is a term used to describe the effect on an object of a force that is acting on that object at a radial distance from the point of interest. The force acts in a way to bend the object around the point of interest.

A force acts in a perpendicular direction at a distance from the point. The moment (M) is a vector quantity. To understand the direction of the moment, we use something called the "Right-Hand-Rule."

Moment: $M = Fd$

Where:
M = moment in Newton-meters (Nm)
F = force in newtons (N)
d = distance in meters (m)

Right Hand Rule

Place your right hand on the table with the fingers pointing straight away from your body and your thumb pointing straight up at a 90° angle from the fingers. Point your fingers in the direction of the distance (d). Then curl your fingers in the direction of the force as if wrapping your fingers around a pole with the thumb pointing up along the pole. Here, the thumb points in the direction of the moment vector.

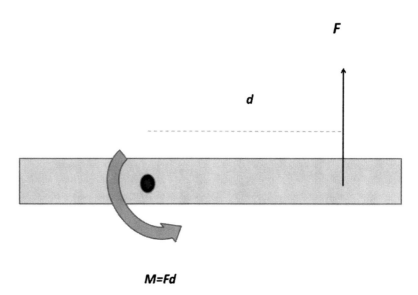

M=Fd

Moment versus Torque

The concept and formula for moment is very similar to that of torque. Moment is a more general examination of the effects of a force applied at a distance. While sometimes this causes a rotation at the axis point, it usually just causes stress. Torque, on the other hand, is a description of the rotational effect of the force applied at a distance and is usually related to a rotation at the axis.

Within mechanical engineering, there is extensive analysis of the various stresses and strains on joints, connections and components of a structure. The bending and flexing of a bridge due to the weight of the bridge itself, the vehicles on the bridge or the wind shear impacts the structural durability of the bridge.

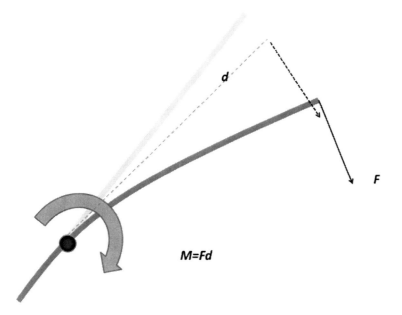

With a cooking pot on the stove filled with water, we lift the pot using the handle. This will cause the pot to experience a moment about the connection point between the handle and the metal pot. This location must be able to endure a lot of stress and is usually reinforced with a metal collar at the connection point between the handle and the pot.

Even a simple computer keyboard experiences moments. If you lift the keyboard by the end with one hand, there is a moment about a point somewhere between your hand and the other end of the keyboard.

Or think of a diving board and how it bends at a particular location along the length of the board. This is the axis around which the moment occurs. This is designed into the board. We can imagine an amount of force at the correct distance that would cause the board to bend the most or even break. These are applications using mechanical engineering analysis to keep people safe, and help them have fun.

A force acts in a direction at a distance from the point. In order to determine the moment (*M*) around a point A, the perpendicular component of the force vector must be determined at a distance d from the axis point A.

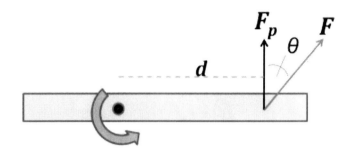

$$M=F_p d = Fd\cos\theta$$

Moment: $M = Fd\sin\theta$

Where:
M = moment in newton-meters (Nm)
F = force in newtons (N)
$F\cos\theta$ = perpendicular component of the force in newtons (N)
d = distance in meters (m)

Example 1: Moment

A force (*F*) of 200 N is applied perpendicular to a rod at a distance (*d*) 55 cm from point A. Determine the moment of the rod about point A.

Using the formula for the moment $M = Fd$ we have:

$M = Fd = 200$ N(55 cm) $= 200$ N(0.55 m) $= 110$ Nm

Example 2: Moment

A force (*F*) of 200 N is applied at a 45°angle to the perpendicular to a rod at a distance (*d*) 55 cm from point A. Determine the moment of the rod about point A.

Using the formula for the moment $M = Fd\cos45°$ we have:

$M = Fd = 200$ N(55 cm)$\cos45° = 200$ N(0.55 m)(0.707) $= 77.78$ Nm

Concept Reinforcement

1. Explain the concept of a moment.

2. Explain the right-hand rule and the direction of the moment.

3. Explain the difference between moment and torque.

4. Calculate the moment about point A of a force of 300 N applied perpendicular to a rod at a distance of 2 m.

5. Calculate the moment about point A of a force of 300 N applied at a 30°angle to the perpendicular a rod at a distance of 2 m.

Section 1.6 – Two-dimensional Force Systems

Section Objective

- Describe two-dimensional force systems

Introduction

This section presents the concept of a set of forces that are acting only in two dimensions, which is a two-dimensional force system. This is easily presented on a sheet of paper. The forces act in any direction, but the forces should be acting on the same object or set of objects, rather than on random objects. This allows the various forces to be added and subtracted either graphically or using calculations involving the Pythagorean theorem and trigonometry.

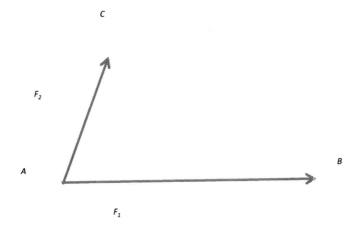

A Force is a Vector

A vector must be shown with both a magnitude and its direction to be complete. A magnitude alone is not sufficient to describe a vector and is then a scalar.

Force is a vector and, therefore, must be presented with both a magnitude and a direction. For instance, describing a force as simple 10 N is insufficient. No direction is defined.

Correct presentations of a force vector:

10 N straight up
20 N to the left
30 N from the right
40 N east

A vector is presented with a magnitude and direction. A positive direction is opposite to a negative direction.

"10 N to the east" is equal in magnitude and opposite in direction to "10 N straight west."

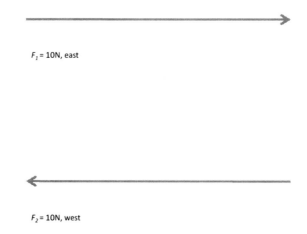

F_1 = 10N, east

F_2 = 10N, west

The magnitude of the vector can also be presented as either positive or negative.

Therefore, "10 N to the east" is equal in magnitude and opposite in direction to "10 N straight east."

The negative sign translates to the same force pointing in the opposite direction. This negative force terminology is applicable, for instance, when a force is applied to slow a vehicle, or decelerate a car.

A single force has a single direction. A two-dimensional force system occurs when a body experiences two forces with directions that are not parallel.

Multiplication of a Vector

Multiplication of a vector means using a scaling factor, a scalar, to increase the size of the vector. The direction of the vector remains the same. The magnitude is changed by the scale of the scalar.

Example 1: Vector Multiplication by a scalar

The force vector of 15 N at an angle of 30° is multiplied by 4.

Define the resultant vector.

Solution

(15 N ∡ 30°) (4)= 60 N ∡ 30°

Two-dimensional force systems

In the illustration, the two forces (F_1, F_2) are in different directions. If both of these force vectors act on the same point (A) or object they can be summed to a single equivalent force vector. The effect of the single force vector F_R acting on the object from point A to point D is exactly the same as the effect of the two force vectors, F_1 acting from A to B and F_2 acting from A to C acting on the object.

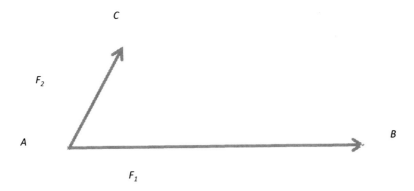

The single equivalent force (F_R) is the "resultant" force vector.

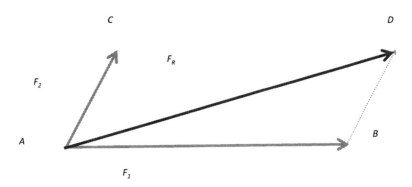

For this to be accurate, the length of the drawn vectors must be properly proportioned. In the end, the resultant force vector length will indicate its magnitude.

Summing Force Vectors Graphically

Vector summation is different from standard algebraic summation. From the illustration, the resultant vector F_R is longer than either of the initial force vectors and the direction is different as well. To sum the vectors graphically we take the second vector and create a parallel vector that starts at the tip of the first vector. In the illustration, that is the dashed vector to the right parallel to F_2. Then the resultant vector is from the original of the first vector to the end of the second.

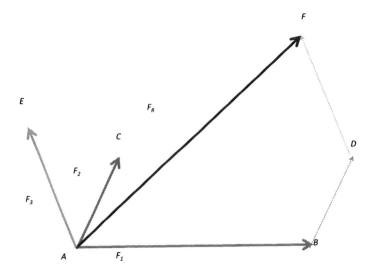

This can easily be expanded to three and more vectors. Graphically we begin with one vector and add each vector in turn starting the subsequent vector from the tip of the previous vector. The effect of the resultant force vector (F_R) will be exactly the same as the effect of the three original force vectors (F_1, F_2, F_3).

Deconstructing Vectors

In a two-dimensional framework, each force vector has an x-component and a y-component. In the illustration, F_1 is shown, along with its x- and y- components.

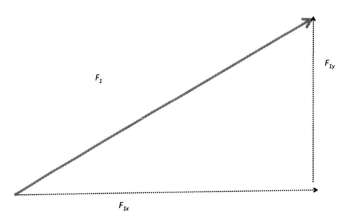

As mentioned above, two vectors are summed by placing the beginning of the second vector at the end of the first vector. The resultant vector F_1 is equivalent to the other two vectors F_{1x} and F_{1y}.

This requires knowledge of the angle of the vector and use of the sine function to determine the y-component and cosine functions to determine the x-component.

For the x-component: $F_{1x} = F_1 \cos\theta$

For the y-component: $F_{1y} = F_1 \sin\theta$

where θ is the angle from the direction of the vector to the horizontal.

The convention used when describing the angle of the vector (θ) is to use directly east (or right) as $0°$. The angle increases in a counterclockwise direction until the angle has made one full circle at which point it is at $360°$ or $0°$. Here the force vector F_{1x} is directly east and at an angle of zero° and the force vector F_{1y} is at a $90°$ angle.

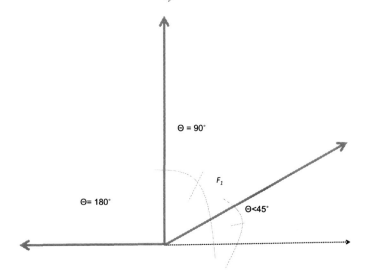

In this way, when several force vectors are to be summed, each of the original vectors can be deconstructed into an x- and a y-component. To determine the resultant vector, sum the x-components together and sum the y-components together and use this as the basis of the x- and y-component of the resultant vector.

Example 1: Summing Force Vectors

We have 3 force vectors; $F_1 = 30$ N at $30°$; $F_2 = 20$ N at $45°$; $F_3 = 60$ N at $60°$.
Deconstruct each vector.
Determine the x- and y-component of each force vector and define the resultant vector.

$F_1 = 30$ N at $30°$
To determine the x-component: $F_{1x} = F_1\cos30°$
$F_{1x} = F_1\cos\theta = 30$ N $(0.866) = 26$ N

To determine the y-component: $F_{1y} = F_1\sin30°$
$F_{1y} = F_1\sin\theta = 30$ N $(0.5) = 15$ N

$F_2 = 20$ N at $45°$;
To determine the x-component: $F_{2x} = F_2\cos45°$
$F_{2x} = F_2\cos\theta = 20$ N $(0.707) = 14$ N

To determine the y-component: $F_{2y} = F_2\sin45°$
$F_{2y} = F_2\sin\theta = 20$ N $(0.707) = 14$ N

$F_3 = 60$ N at $60°$;
To determine the x-component: $F_{3x} = F_3\cos60°$
$F_{3x} = F_3\cos\theta = 60$ N $(0.5) = 30$ N

To determine the y-component: $F_{3y} = F_3 \sin 60°$

$F_{3y} = F_3 \sin\theta = 60 \text{ N } (0.866) = 52 \text{ N}$

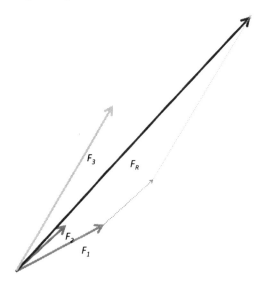

F_R; resultant force

Sum the x-components.

$F_{1x} + F_{2x} + F_{3x} = 26 \text{ N} + 14 \text{ N} + 30 \text{ N} = 70 \text{ N}$

Sum the y-components.

$F_{1y} + F_{2y} + F_{3y} = 15 \text{ N} + 14 \text{ N} + 52 \text{ N} = 81 \text{ N}$

The magnitude of the resultant vector is obtained using the Pythagorean theorem.

$$F_R = \sqrt{F_{Rx}^2 + F_{Ry}^2} = \sqrt{(70 \text{N})^2 + (81 \text{N})^2} = 107 \text{N}$$

The direction is obtained using the arctangent function:

$$a\tan\frac{81}{70} = 49°$$

The resultant force vector is: $F_R = 107 \text{ N}, 49°$.

Example 2: Summing Force Vectors

In this example, we have 3 force vectors; $F_1 = 30 \text{ N}$ at $30°$; $F_2 = 20 \text{ N}$ at $45°$; $F_3 = 40 \text{ N}$ at $150°$.

Deconstruct each vector.

Determine the x- and y-component of each force vector and define the resultant vector.

$F_1 = 30 \text{ N}$ at $30°$;

To determine the x-component: $F_{1x} = F_1 \cos 30°$

$F_{1x} = F_1 \cos\theta = 30 \text{ N } (0.866) = 26 \text{ N}$

To determine the y-component: $F_{1y} = F_1 \sin 30°$

$F_{1y} = F_1 \sin\theta = 30 \text{ N } (0.5) = 15 \text{ N}$

F_2 = 20 N at 45°;
To determine the *x*-component: $F_{2x} = F_2\cos45°$
$F_{2x} = F_2\cos\theta = 20$ N (0.707) = 14 N

To determine the *y*-component: $F_{2y} = F_2\sin45°$
$F_{2y} = F_2\cos\theta = 20$ N (0.707) = 14 N

F_3 = 40 N at 150°;
To determine the *x*-component: $F_{3x} = F_3\cos150°$
$F_{3x} = F_3\cos\theta = 40$ N (-0.866) = -35 N

To determine the *y*-component: $F_{3y} = F_3\sin150°$
$F_{3y} = F_3\sin\theta = 40$ N (0.5) = 20 N

F_R; resultant force
Sum the *x*-components.
$F_{Rx} = F_{1x} + F_{2x} + F_{3x} = 26$ N + 14 N + (-35 N) = 5 N

Sum the *y*-components.
$F_{Ry} = F_{1y} + F_{2y} + F_{3y} = 15$ N + 14 N + 20 N = 49 N

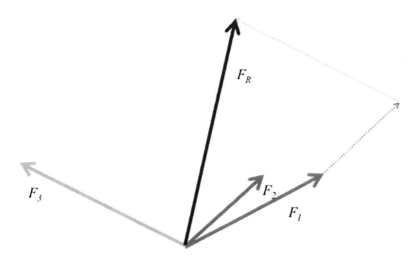

The magnitude of the resultant vector is calculated with the Pythagorean theorem.

$$F_R = \sqrt{F_{Rx}^2 + F_{Ry}^2} = \sqrt{(5\text{N})^2 + (49\text{N})^2} = 49\text{N}$$

The direction is obtained using the arctangent function:

$$a\tan\frac{49}{5} = 84°$$

The resultant force vector is: F_R = 49 N, 84°.

Concept Reinforcement

1. What is required to fully describe a force vector.

2. Explain what a negative magnitude of a vector indicates.

3. Describe how to graphically add two vectors.

4. With the 3 force vectors; $F_1 = 10$ N at $30°$; $F_2 = 20$ N at $45°$; $F_3 = 50$ N at $135°$, deconstruct each vector. Determine the x- and y-component of each force vector and define the resultant vector.

Section 1.7 – Three-dimensional Force Systems

Section Objective

- Describe three-dimensional force systems

Introduction

This section presents the concept of a force system that exists in three dimensions. This is closer to reality than the two-dimensional force systems. Most objects in reality experience forces in all three dimensions. This is, of course, more complicated, but using x, y, z coordinates simplifies the calculation and the understanding of the methods to arrive at the resultant force vector. Each force has an x-component, a y-component and a z-component. The resultant force vector is the summation of all the x-components, y-components and z-components.

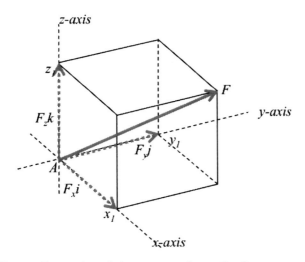

Three-dimensional deconstruction of a force vector

A Force is a Vector

For a complete description a force vector must be presented as a magnitude along with its direction. A magnitude alone is insufficient information for describing a vector.

Correct presentations of a force vector:

> 105 N straight down
> 50 lb at an angle of 20°
> 20 N to the left which is equivalent to 20 N to the right

Three-dimensional force systems

Forces on mechanical structures act in three dimensional space. This is a more accurate representation of the actual conditions in the world of engineering work. This is an extension of the analysis of force vectors using two-dimensional space and will be utilized often in the analysis of the more interesting advanced aspects of mechanical engineering.

In the illustration, the single force (F) is deconstructed into its x, y, and z components, (F_x, F_y, F_z) along each axis of the Cartesian coordinate system.

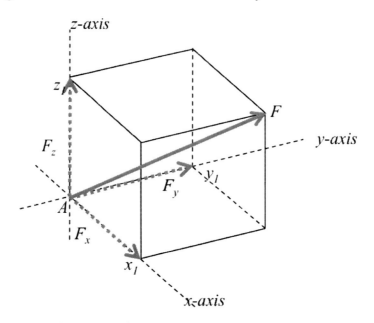

3-dimensional deconstruction of a force vector

The effect on the point (A) of the summation of these three forces (F_x, F_y, F_z) is equivalent to the effect on the point (A) of the single force (F).

Addition of 3-D vectors

When several 3D vectors are to be added, the x, y, and z coordinates are summed individually. This assumes that all vectors begin at the origin (0, 0, 0). With vectors 1, 2, 3…n, the resultant vector will be defined by the three coordinates: $R_x = 1_x + 2_x + 3_x + …n_x$; $R_y = 1_y + 2_y + 3_y + …n_y$; $R_z = 1_z + 2_z + 3_z + …n_z$.

Example 2: 3D vector addition

Given two vectors starting at the origin (0, 0, 0), Vector 1 with coordinates (3, 4, 5) and Vector 2 with coordinates (6, 7, 8), calculate the resultant vector as a result of adding these two vectors.

Solution

Resultant = Vector 1 + Vector 2 = (3, 4, 5) + (6, 7, 8) = (3+6, 4+7, 5+8) = (9, 11, 13).

The Cartesian basis vector

The Cartesian basis vector (i, j, k) is the identity vector with the coordinates of (1, 1, 1) and beginning at the origin.

i is the vector from (0, 0, 0) to (1, 0, 0)

j is the vector from (0, 0, 0) to (0, 1, 0)

k is the vector from (0, 0, 0) to (0, 0, 1)

This allows any vector to be defined as a linear combination of the basis vector.

The illustration shows the vector (F) and its x, y, and z components, (F_x, F_y, F_z).

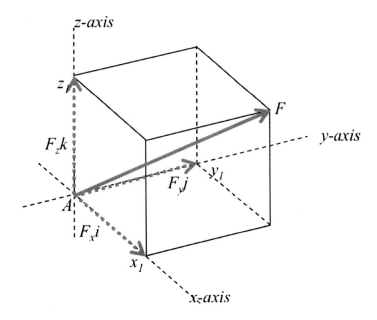

Three-dimensional deconstruction of a force vector

Here the vector can be defined as:

$$F = F_x i + F_y j + F_z k.$$

Deconstructing 3-Dimensional Force Vectors

The magnitude of each of the deconstructed force vector components is determined by the angle on the force vector (F) with respect to that axis.

Magnitude of the x-component: $F_x = F\cos\alpha$

Where:
F_x = the vector along the x-axis of with magnitude from point A to point x_1.
α = the angle from the x-axis to the force vector.

Magnitude of the y-component: $F_y = F\cos\beta$

Where:
F_y = the vector along the y-axis of with magnitude from point A to point y_1.
β = the angle from the y-axis to the force vector.

Magnitude of the z-component: $F_z = F\cos\gamma$

Where:
F_z = the vector along the z-axis of with magnitude from point A to point z_1.
γ = the angle from the z-axis to the force vector.

The Pythagorean Theorem used to solve two-dimensional systems is also applicable to 3-dimensional space.

$$F_R = \sqrt{F_x^2 + F_y^2 + F_z^2}$$

In the illustration the x-component of the force vector on the x-axis is presented with the angle (in the shaded plane) from the vector to the x-axis being α.

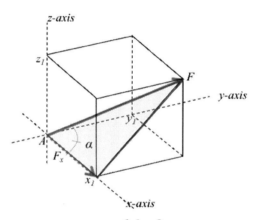

x-component of the force vector

A similar projection can be produced for all three components and angles; the x-component, the y-component and the z-component. In all cases, the angles will be as listed above.

This vector deconstruction in 3-dimensions uses the same general concepts as two-dimensional vector deconstructions.

Summing Force Vectors Graphically

The force vector (i.e. the resultant vector) can be developed graphically from the component vectors by simply starting with one component vector and then adding the subsequent component vectors starting at the tip of the previous component vector. This is done for all three component vectors. The origin of the first component vector is the origin of the resultant vector and the tip of the last component vector is the tip of the resultant vector.

The third illustration shows the graphical summation. Clearly, the *y*-component vector has just been moved forward and the *z*-component vector has been moved diagonally to the opposite corner.

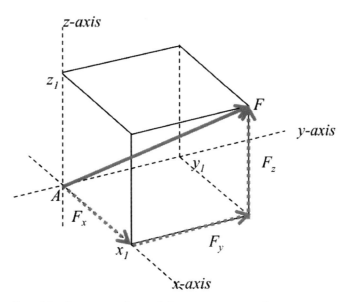

Graphical summation of the component force vectors

For this to be accurate, the length of the drawn force vectors must be properly proportioned. In the end, the resultant force vector length will indicate its magnitude.

Example 1: 3-Dimensional Vector Summation

We have a force vector of 7.07 N in 3-dimensional space that begins at the origin $(0, 0, 0)$ and ends at the coordinates $(x, y, z) = (3, 4, 5)$. Deconstruct the vector and determine the angles between the *x*, *y* and *z* component and the force vector. State the resultant vector using the basis vector.

Solution

A: *x*-component angle:

The magnitude of the *x*-component (3) is defined by: $F_x = F\cos\alpha$

First it's necessary to determine the angle (α) between the force vector and the *x*-axis. This is the angle projected in the plane defined by the resultant vector and the *x*-axis.

From $F_x = F\cos\alpha$ and solving for the cosine of the angle: $\cos\alpha = \dfrac{F_x}{F}$ and then the angle itself: $\alpha = \arccos\left(\dfrac{F_x}{F}\right)$ and $\alpha = \arccos\left(\dfrac{F_x}{F}\right) = \arccos\left(\dfrac{3}{7.07}\right) = 64.9°$

The magnitude of the *y*-component (4) is defined by: $F_y = F\cos\beta$

Again, it's necessary to determine the angle (β) between the force vector and the *y*-axis. This is the angle projected in the plane defined by the resultant vector and the *y*-axis.

From $F_y = F\cos\beta$ and solving for the cosine of the angle: $\cos\beta = \dfrac{F_y}{F}$ and then the angle

itself: $\beta = \arccos\left(\dfrac{F_y}{F}\right)$ and $\beta = \arccos\left(\dfrac{F_y}{F}\right) = \arccos\left(\dfrac{4}{7.07}\right) = 55.5°$

C: *z*-component angle:

The magnitude of the *z*-component (5) is defined by: $F_z = F\cos\gamma$

Again, it's necessary to determine the angle (γ) between the force vector and the *z*-axis. This is the angle projected in the plane defined by the resultant vector and the *z*-axis.

From $F_z = F\cos\gamma$ and solving for the cosine of the angle: $\cos\gamma = \dfrac{F_z}{F}$ and then the angle

itself: $\gamma = \arccos\left(\dfrac{F_z}{F}\right)$ and $\gamma = \arccos\left(\dfrac{F_z}{F}\right) = \arccos\left(\dfrac{5}{7.07}\right) = 45°$

To state the resultant vector using the basis vector we use the model of:

$$F = F_x i + F_y j + F_z k.$$

Here the resultant vector is: $F_R = 3i + 4j + 5k.$

Concept Reinforcement

1. Explain how to present a vector in 3-dimensions.

2. Add the three 3D vectors (3, 6, 9), (2, 6, 8) and (1, 2, 3). Assume they all begin at the origin (0, 0, 0).

3. Express the vector (7, 2, 9) using the basis vector.

4. Determine the angles between the *x*, *y*, and *z* components and the resultant vector in the basis vector.

Section 1.8 – Moments and Moment Vectors

Section Objective

- Discuss moments and moment vectors

Introduction

This section presents mechanical moments and the moment vector. This introduces the right-hand rule as a guide to determine the actual direction of the moment vector. A moment is a way to quantify the combined effect of a force at a radial distances from a point. A larger force increases the moment and a larger distance increases the moment. Generally, the direction of the moment is perpendicular to the direction of the force. It is also perpendicular to the direction of the radial distance between the force and the point of interest. The moment can be described as rotating in a clockwise or counterclockwise direction, though this loses its relevance when moments occur in various directions in 3-dimensional space. Therefore, it's important to use a vector arrow in illustrations that shows the direction and magnitude of the moment.

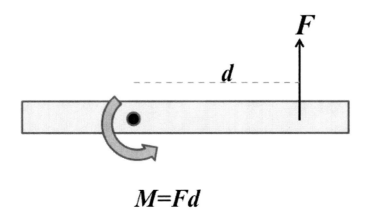

$$M=Fd$$

Moment Description

Moment is a term used to describe the effect of a force acting on a rigid object at a radial distance from the point of interest. The force acts in a way to bend the object around the point of interest. As an example, think of a diving board and how it moves when the diver is ready to jump from the board.

The force may be acting at any conceivable angle, but the moment is produced by the component of the force that acts perpendicular to the radial line connecting the force and the point. The moment (M) is a vector quantity and to understand the direction of the moment we use something called the "Right-Hand-Rule."

Moment: $M = Fd$

Where:
M = moment in newton-meters (Nm)
F = force in newtons (N)
d = distance in meters (m)

Right Hand Rule

Place your right hand on the table with the fingers pointing straight away from your body and your thumb pointing straight up at a 90° angle from the fingers. Point your fingers in the direction of the distance (d). Then curl your fingers in the direction of the force as if wrapping your fingers around a pole with the thumb pointing up along the pole. At this point, the thumb points in the direction of the moment vector.

Rigid Body Assumption

To simplify the analysis at this stage all objects are assumed to be rigid bodies. This means that there is no bending of the object and the effects of the forces and moments are felt equally throughout the object. Contrast this to an object that is flexible or pliable. A force experienced at one end may not be experienced with the same intensity at the other end. This would complicate the analysis and, of course, is required at more advanced stages of mechanical engineering work. For now all objects will only be considered rigid bodies.

Line of Action of a Force

All forces are vectors and act along a direction called the line of action. In the illustration, the force is shown in red and the line of action is shown as a dashed line. The object experiences the force as if it is acting anywhere along the line of action.

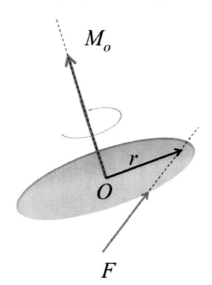

Demonstrating the line of action of the force

Force and Moment

A force can be in any direction, therefore a moment can also be in any direction. Any object subjected to a force has the potential to also experience a moment at different points on the body. This discussion begins with examining objects in two dimensions and expands to 3-dimensional space in a Cartesian coordinate system.

Moment Vectors

As presented in the explanation of the right-hand rule, the direction of the moment is perpendicular to the plane of the object in which the radial distance is measured. The moment is also perpendicular to the perpendicular component of the force applied at the radial distance from the point of interest.

In the illustration, the force is acting along its line of action, which is not perpendicular to the radial distance (r) as shown.

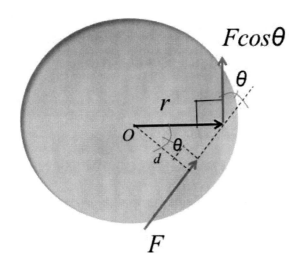

A top down view of the plane

From the formula for the moment, it is possible to establish the magnitude of the moment in two ways:

1. using the fraction of the force that is perpendicular to the radial distance (r).

2. using the fact that this is a rigid body and establishing a distance (d) that is perpendicular to the line of action of the force.

In both cases, the formula will result in the same value for the moment about the point O.

Using the perpendicular component of the force ($F\cos\theta$), the formula for the moment is:

Moment $= (F\cos\theta)r = Fr\cos\theta$.

Using the distance (d) that is perpendicular to the line of action of the force, we have

$d = r\cos\theta$ and the moment is:

Moment = $Fd = F(r\cos\theta) = Fr\cos\theta$.

Example 1: Moment Vector

Using the illustrations in the section above, we have a radial distance of 50cm and a force of 200 N acting on a line of action that is 30° from the perpendicular to the radial distance where the force and the radial distance meet.

Calculate the moment using the two different formulas and describe the direction of the moment.

First determine the cos30° which is 0.866.

Using the component of the force that is perpendicular to the radial distance:

$F\cos30° = 200 \text{ N } (0.866) = 173$ N.

Moment = $(F\cos\theta)r = Fr\cos\theta = 173 \ (0.5 \text{ m}) = 86.6$ Nm

Using the distance (d) that is perpendicular to the line of action of the force,

$d = r\cos30° = 0.5 \text{ m } (0.866) = 0.433$ m.

Moment = $Fd = F(r\cos\theta) = Fr\cos\theta = 200 \text{ N } (0.433\text{m}) = 86.6$ Nm

In both cases, the moment is perpendicular to the plane described by the line of action of the force and the radial distance. The direction is counterclockwise.

Concept Reinforcement

1. Explain the right-hand rule for moments.

2. Explain the rigid body assumption.

3. Explain the line-of-action of a force.

4. Using the same illustrations from the section above, define the moment using the two different methods. The radial distance (r) is 70 cm, the force (F) is 300 N and the angle of approach of the line-of-action is 60° to the perpendicular to the radial distance.

Section 1.9 – Couples

Section Objective

- Describe couples and their uses in engineering

Introduction

This section presents couples, which are pairs of forces that are equal in magnitude, parallel and opposite in direction. These force couples cause the object to experience a moment that has a direction defined by the right hand rule. The magnitude of the moment is defined by the equation $M = Fd$. An object can experience a force couple at any conceivable angle, which in turn means that the moments which are directed perpendicular to the plane of the force couple are also in any conceivable direction.

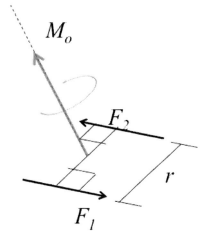

A force couple

Couple Definition

A couple is a pair of forces that are parallel, but in opposite directions. This pair of forces creates a moment about a point midway between the parallel lines of action of the forces.

Each force is equal and each is $r/2$ from the center point and therefore the two moments $M_1 = F_1\dfrac{r}{2}$ and $M_2 = F_2\dfrac{r}{2}$ result in a single moment of: $M_o = F_1\dfrac{r}{2} + F_2\dfrac{r}{2} = 2F\dfrac{r}{2} = Fr$

These forces can be paired in any conceivable direction and, therefore, the moments that are products will be possible in any conceivable direction, as well.

In the illustration, the forces are presented perpendicular to the radial distance. It is just as possible to define a radial distance at an angle to both forces. The magnitude of the moment will still be equal to the perpendicular distance between the forces. The perpendicular distance will be determined as a function of the angle.

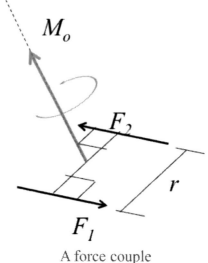

A force couple

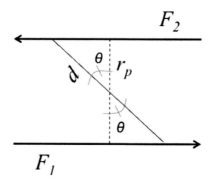

Two parallel forces (a force couple) with the line-of-action of the two forces

The moment will also be located in the center between the parallel lines of action of the two forces. The radial distance that is perpendicular to the two force vectors (r_p) is determined through the formula:

Radial distance: $r_p = d\cos\theta$

The moment is: $M = Fr_p$

Recall that each force-distance pair is ½ the total, which is why the moment formula is equal to the total perpendicular separation distance times one force magnitude.

It is also important to understand that the location of the moment is determined by the location of the parallel lines of the line-of-action of each of the forces. In the illustration there are two parallel forces acting where their lines-of-action are parallel but the two forces appear disconnected.

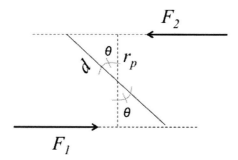

Two parallel forces (a force couple) with the line-of-action of the two forces

Couples in 3-Dimensions

Forces act in all directions and when parallel forces act this creates a force couple.

The illustration presents two pairs of forces (two force couples) acting in different directions on a 3-dimensional rigid body. This is more common than the two dimensional pattern.

In most structures, there are many moments acting at the same time due to force couples.

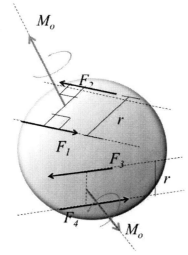

Two paired parallel force couples with the line-of-action
of the two forces visible and the moments visible

Force and Couple System

When a set of paired forces and additional forces are present, the paired forces represent a couple producing a moment and the forces have a separate effect.

These separate forces act on the body in ways different from the force couples.

These separate forces also create moments separate from the force couples.

Concept Reinforcement

1. Describe a couple.

2. State the formula for the moment that is created by a couple.

3. Explain the direction of the moment created by a couple.

Section 1.10 – Equivalent Systems

Section Objective

- Explain equivalent systems

Introduction

This section presents the concept of equivalent systems. These systems include only external forces, couples and the associated moments generated by the force couples. When the net effect on the object is the same with two apparently different sets of forces, couples and moments then the two systems are considered equivalent. Technically, the net effect of all the forces must be equivalent and the net effect of all the moments must be equivalent, in order for the two systems to be equivalent.

Equivalent Systems

When the resultant forces of two systems are equivalent and the resultant moments are equivalent then the two systems are equivalent. This can be approached in two ways; either we can examine two random systems to determine if they are equivalent or we can change a known system to an equivalent system.

Here we focus on the second approach where a known system is changed to an equivalent system. There are advantages to changing the location of a force and relocating the moment of the forces. There are three general changes; moving a force along its line of action, moving a force off its line of action and define the resultant of a force and couple system.

Moving a force along its line of action

Moving a force along its line of action creates a new force system. This new system is equivalent to the original force system.

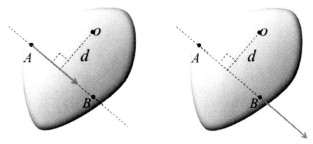

These two force systems are equivalent.
The force has been moved along its line of action.

Note that the force creates a moment about the point O and that the radial distance to the line of action is the same in both systems. Therefore the direction of the force and the direction of the radial distance are both the same. Since the magnitudes of these two aspects are equivalent in the two systems the moments are equivalent.

Moving a force off its line of action

When moving a force off its line of action, a couple must be added to the new force system in the new location to compensate for the change.

Here the original force system includes a force acting on point A. The objective is to produce an equivalent force system with an equivalent force at point B.

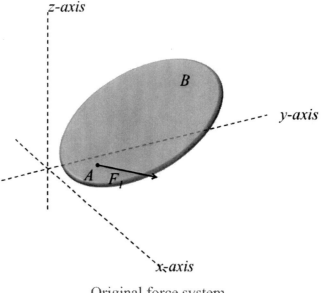

Original force system

The second stage includes a couple which is created by adding a positive force and a negative force at point B both parallel to and the same magnitude as the original force at point A.

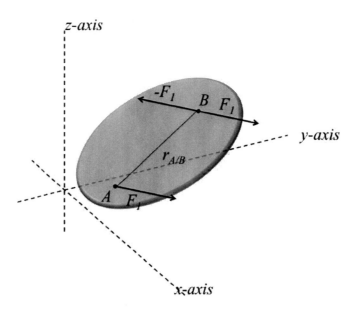

The couple added as part of the move off the line of action of the original force

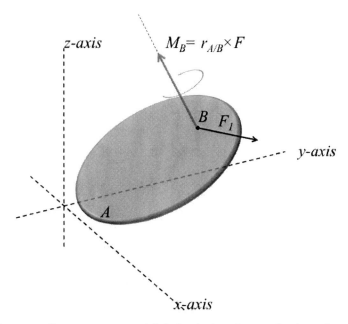

The new force system, which includes the equivalent force
at point B and a moment at point B

The final stage changes the couple (F_1 at point A and $-F_1$ at point B) to a moment (M_B) at point B. This is the moment that would be experienced at point B if the force had stayed at point A.

Note: All three of these systems are equivalent.

The resultant of a force and couple system

At any point O every force and couple system can be modified to be equivalent to a single force passing through the point O and a single couple. The single force (F_R) is equal to the resultant force of the original system. The single couple (C_R) is equal to the resultant moment of the original system around the point O.

The first illustration includes two couples (red) and four forces at varying distances from the point of interest (point O).

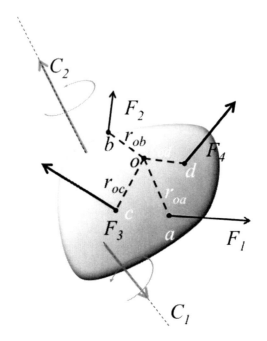

Original force system

In the second illustration, all the forces have been shifted off their line of action to the point O. Each shift includes a couple.

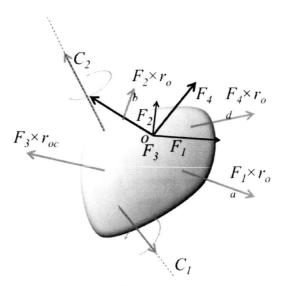

Here all the forces have been moved off their line of action and are now starting at the point O and include a couple (F_r) associated with each force.

The final force system is one single force and one single couple.

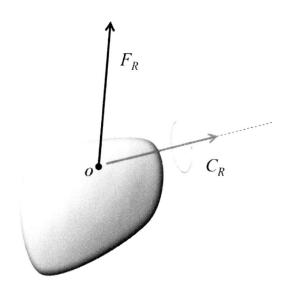

The final force system includes a single force starting
at the point O and a couple that starts at point O.

All three of these systems are equivalent.

Concept Reinforcement

1. Explain the concept of equivalent systems

2. Explain the idea of a force moving along its line of action.

3. Explain how to move a force from its line of action.

4. Explain the concept of the resultant of a force and couple system

Section 1.11 – Systems of Algebraic Equations

Section Objective

- Explain how to solve systems of algebraic equations

Introduction

This section presents the concept of a system of algebraic equations, which is a set of equations that are assumed to have at least one point in common. These "systems" of equations can be solved using graphical means by drawing the lines described by the equations on a graph. These sets of equations can also be solved using algebra. This is done by either adding or subtracting one equation to or from the other. The system can also be solved by solving one of the equations for one of the variables and substituting the result into the other equation.

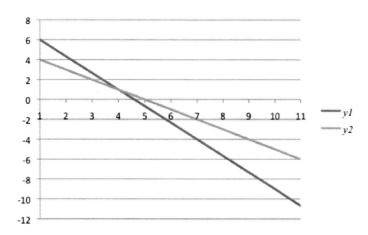

Systems of Simultaneous Equations

When there are two straight lines in a plane that intersect, the point of intersection is a coordinate (x, y), which is in the set of solutions for both of these lines. This can be extended to solve three, four and more simultaneous lines.

The solution of this coordinate (x, y) can be determined using graphical means or through solving algebraic equations such as $x + 2y = 1$.

In algebra, when one equation is given and a single variable is unknown, the solution is straightforward.

Example 1: Equation with one unknown variable

Given the equation: $5 + 3y = 23$, solve for y.

Solution

This is a simple matter of subtracting 5 from both side to arrive at the equation: $3y = 18$. From this both sides are divide by 3 to arrive at $y = 6$, which is the solution.

It is therefore necessary to have one equation in order to solve for one unknown.

Also, in algebra, when there are two unknowns, it is necessary to have two equations.

Example 2: Single Equation with two unknown variables

Given the equation: $5x + 3y = 23$, solve for x and y.

Solution

With just a single equation, we can prove that for every x there is a corresponding y. There is no unique solution because there is only the single equation.

This section presents three different approaches to solving the problem of two unknowns. These methods are applicable to linear as well as non-linear equations. These methods can be extended to any number of unknowns, though computationally, the work becomes quite complicated with many unknowns.

The three methods are:

- a graphical approach

- using the equations

- solving algebraically by addition and by substitution.

Graphical Solutions

The graphical method for solving the problem of the unique (x, y) coordinate that works for both lines is simply drawing each line on a graph and evaluating where they intersect. The intersection is the point at which the (x, y) coordinate is in the set of solutions for both of the lines.

Example 3: Two equations with two unknowns

Given the equation: $5x + 3y_1 = 23$ and $x + y_2 = 5$, solve for the one unique coordinate pair for x and y that is a solution for both lines.

Solution

Both equations are drawn in the figure. The *x*-value is along the *x*-axis. The two lines appear to intersect at a position where $x = 4$.

This can be verified using the two equations and solving for the y values.

$5x + 3y_1 = 23$ and substituting $x = 4$ we have $20 + 3y_1 = 23$. Subtracting 20 from both sides, we have $3y_1 = 3$ and $y_1 = 1$.

$x + y_2 = 5$ and substituting $x = 4$ we have $4 + y_2 = 5$. Subtracting 4 from both sides we have: $y_2 = 1$.

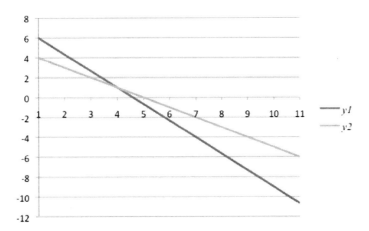

Therefore, both equations have the same pair of coordinates (4, 1). However, this can only be determined accurately through algebraic means. Using graphical means, there will always be a larger margin of error in the solution as compared to the algebraic methods.

Linear System of Algebraic Equations

When the lines are described by linear equations, it is called a linear system of equations.

Solving through Elimination by Addition

When solving for the variables in a set of equations the primary goal is to eliminate one of the variables at a time. This can be done by manipulating one or both equations, and then using addition (or subtraction) of the two equations to remove one of the variables.

Example 4: Two equations with two unknowns

Given the equation: $5x + 3y = 23$ and $x + y = 5$, solve for the one unique coordinates for *x* and *y* that is a solution for both lines.

Initially we have the two equations:
$$5x + 3y = 23$$
$$x + y = 5$$

We need to eliminate one of the two variables, either x or y.

We will solve this first by eliminating x.

We modify the second equation:
$$\begin{aligned} 5x + 3y &= 23 \\ 5(x + y &= 5) \end{aligned}$$
which becomes
$$\begin{aligned} 5x + 3y &= 23 \\ 5x + 5y &= 25 \end{aligned}$$

We eliminate the variable x by subtracting the second equation from the first:

$$\begin{aligned} 5x + 3y &= 23 \\ \underline{5x + 5y} &= \underline{25} \\ -2y &= -2 \end{aligned}$$
and from this, we have a single equation and a single unknown and the

solution is $y = 1$. This is then replaced in either equation to solve for the other unknown, x.

Using the equation $x + y = 5$, we substitute the known value for y ($y = 1$) and obtain: $x + 1 = 5$. Subtracting 1 from each side results in $x = 4$.

This can also be solved by eliminating the y variable.

We modify the second equation:
$$\begin{aligned} 5x + 3y &= 23 \\ -3(x + y &= 5) \end{aligned}$$
which becomes
$$\begin{aligned} 5x + 3y &= 23 \\ -3x - 3y &= -15 \end{aligned}$$

We eliminate the variable y by adding the second equation to the first:
$$\begin{aligned} 5x + 3y &= 23 \\ \underline{-3x - 3y} &= \underline{-15} \\ 2x &= 8 \end{aligned}$$

and from this we have a single equation and a single unknown and the solution is $x = 4$. This is then replaced in either equation to solve for the other unknown, y.

Using the equation $x + y = 5$ we substitute the known value for x ($x = 4$) and obtain: $4 + y = 5$. Subtracting 4 from each side results in $y = 1$.

Note that if both lines have the same slope there is no solution because the lines do not intersect anywhere.

Solving through Elimination by Substitution

Again, when solving for the variables in a set of equations, the primary goal is to eliminate one of the variables at a time. This can also be done by using one of the equations to develop a substitution, where one of the variables is defined in terms of the other variable and a constant. This is used to replace the variable in the other equation so that there is only a single equation with a single unknown which is easily solvable.

Example 4: Two equations with two unknowns

Given the equation: $5x + 3y_1 = 23$ and $x + y_2 = 5$, solve for the one unique coordinates for x and y that is a solution for both lines.

Initially we have the two equations:
$$5x + 3y = 23$$
$$x + y = 5$$

We need to create a substitute for one of the two variables, either x or y.

We will solve this first creating a substitute for x.

We modify the second equation: $x = 5 - y$, which is then substituted into the first equation, which becomes $5(5 - y) + 3y = 23$ and $25 - 5y + 3y = 23$ and finally, $-2y = -2$ which results in $y = 1$.

This is then substituted into the first equation $x + y = 5$ to obtain: $x + 1 = 5$. Subtracting 1 from each side results in $x = 4$.

This can also be solved by creating a substitute for y.

We modify the second equation: $y = 5 - x$ which is then substituted into the first equation which becomes $5x + 3(5-x) = 23$ and $5x + 15 - 3x = 23$ and finally $2x = 8$, which results in $x = 4$.

This is then substituted into the first equation $x + y = 5$ to obtain: $4 + y = 5$. Subtracting 4 from each side results in $y = 1$.

Concept Reinforcement

1. Describe a system of linear algebraic equations

2. Describe the graphical method of solving a system of two equations.

3. Explain solving through elimination by addition.

4. Explain solving through elimination by substitution.

Section 1.12 – Two-Dimensional Equilibrium

Section Objective

- Explain how to solve two-dimensional equilibrium problems

Introduction

This section presents the concept of a two-dimensional equilibrium problem. Equilibrium is a state where the forces and the moments acting on an object all sum to zero. Solving these problems therefore requires that the forces in both dimensions, both the horizontal (x) axis and the vertical (y) axis sum to zero and the moments, which have a direction perpendicular to the plane and are therefore in the z-axis, also sum to zero.

Equilibrium

Mechanical equilibrium is a condition of an object where the sum of the forces acting on the object is zero and the sum of the moments acting on the object is zero.

Mechanical Equilibrium: $\sum F = 0$ and $\sum M = 0$

When considering two-dimensional equilibrium we have:

Mechanical Equilibrium: $\sum F_x = 0$, $\sum F_y = 0$ and $\sum M_z = 0$

The vector sum of the external forces and force components in the x-direction is zero and the sum of the external forces and force components in the y-direction is zero.

The sum of the moments acting on the object in the z-direction is zero. All moments in a two-dimensional (x, y) system act in the z-direction.

Because of the assumption that the object is a rigid body, this definition can be expanded.

Mechanical equilibrium is a condition of an object where the vector sum of the forces acting on all the particles of the object is zero and the sum of the moments acting on all the particles of the object is zero.

Newton's Laws of Motion

Isaac Newton developed several laws of motion that relate to the issue of forces and equilibrium.

Newton's First Law of Motion:

A particle that is at rest remains at rest unless acted upon by an outside force.

A particle that is moving with a constant velocity continues to move with constant velocity unless acted upon by an outside force.

These statements both assume that the outside force is a single force not countered by an equal and opposite force.

The first law supports the concept that an object in equilibrium must have zero resultant force and zero moment acting on it.

Newton's Second Law of Motion:

The force (F) applied to an object of mass (m) causes acceleration (a) of the object in the direction of the force vector and the relationship is: $F = ma$.

The second law explains that if a force is applied, the object must be accelerated.

Newton's Third Law of Motion:

For every action force of object X on object Y, there is a reaction force of object Y on object X, which is equal in magnitude, opposite in direction and acts at exactly the same time.

The third law explains that to have an object in mechanical equilibrium requires equal and opposite forces between the object and its environment.

Zero point for Angles

In order to maintain a consistent treatment for forces and moments, the angles will be defined from the horizontal vector pointing to the right. This is established as zero degrees (0°) and the magnitude of the angle increases in the counterclockwise direction until the angle of the vector is pointing again directly to the right. At this point the vector has traversed one full rotation and is at an angle of 360° which is equivalent to 0°.

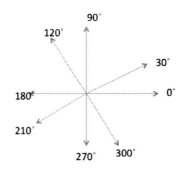

Two-Dimensional Equilibrium

This section focuses on equilibrium in a two-dimensional system in the x-y plane. Forces can act in the x or y directions or any angular direction in the x-y plane. All moments will therefore be directed in the z-direction. In a system in equilibrium, all the forces and force components sum to zero and all the moments sum to zero.

To solve these equilibrium problems we need to identify the forces and force components acting in the x-direction and those acting in the y-direction. The objective is to determine the amount of force in each direction, which will results in mechanical equilibrium where the sum of the forces and the moments is zero:

Mechanical Equilibrium: $\sum F_x = 0$, $\sum F_y = 0$ and $\sum M_z = 0$. For forces that are directed along either the x or the y axis the sum of the force is clearly just the magnitude of the force. For forces that are directed at an angle from the horizontal, the determination of the x and y components requires trigonometry.

Two-Dimensional Equilibrium Problems

Example 1: 2D Equilibrium

A 4 m long bar with a mass of 200 kg is held by two equidistant supports. The system is in equilibrium. Determine the force on the two supports. Determine the moments acting within the system.

Solution

First we develop an equation that describes this system.

Because the system is in equilibrium, the forces supplied by the supports must be in the up direction to counter the force of gravity acting on the mass. Therefore, the system can be described as:

$$F_1 + F_2 = 200 \text{ kg}(9.8 \text{ m/s}^2)$$

Note that the acceleration of gravity is 9.8m/s² and the force in the down direction of the 200 kg mass is $F = 200 \text{ kg}(9.8 \text{ m/s}^2)$.

We know that the two supports are equidistant and the object is treated as a rigid body, therefore the forces are equivalent ($F_1 = F_2 = F_s$) where F_s is the support force.

The equation can be simplified to $2F_s = 200 \text{ kg}(9.8 \text{ m/s}^2)$.

The solution is to divide both sides by 2 to obtain: $F_s = \dfrac{200 \text{ kg}}{2} 9.8 \text{ m/s}^2$

Finally the force on each support is: $F_s = (100 \text{ kg}) (9.8 \text{ m/s}^2) = 980 \text{ N}$

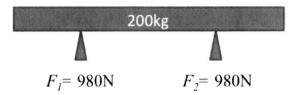

$F_1 = 980\text{N}$ $F_2 = 980\text{N}$

Because the bar has its mass evenly distributed, there is no moment produced by the bar itself.

Types of Two-Dimensional Equilibrium Problems

There are four general types of two-dimensional equilibrium problems. These are the collinear force system, the parallel force system, the concurrent force system and the general force system.

Collinear Force System

In a collinear force system all the forces are acting on the exact same line of action. The illustration presents how this might appear. With a multitude of forces all acting along the same line of action the forces must sum to zero for equilibrium to exist.

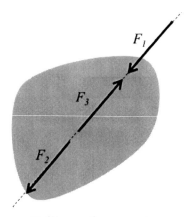

Collinear force system

$$\sum_{i=1}^{n} F_i = 0$$

In the collinear force system there are no moments.

Parallel Force System

In a parallel force system, all the forces act on parallel lines of action. The illustration presents how this might appear. With a multitude of forces all acting along parallel lines of action, the forces must sum to zero and the moments must sum to zero for equilibrium to exist.

When the forces are parallel to one particular axis, such as the x-axis the requirements for equilibrium reduce to:

$$\sum F_x = 0 \qquad \sum M_z = 0$$

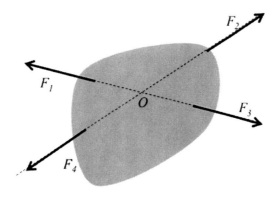

Parallel force system

$$\sum F_x = 0 \qquad \sum F_y = 0 \qquad \sum M_z = 0$$

Concurrent Force System

In a concurrent force system all the forces are acting from the exact same origin. The illustration presents how this might appear. With a multitude of forces all acting along the same line of action the forces must sum to zero for equilibrium to exist.

Concurrent force system

$$\sum F_x = 0 \qquad \sum F_y = 0$$

This system, because all the forces are acting from the same point, has no moments.

In a general force system, all the forces are acting in random directions on the object and produce various moments. The illustration presents how this might appear. With a multitude of forces all acting along different lines of action the vector sum of the forces and force components must equal zero for equilibrium to exist. Also, in this case the moments about the *z*-axis must sum to zero for equilibrium to exist.

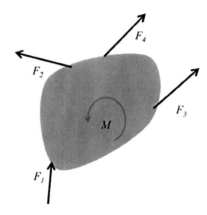

General force system

$$\sum F_x = 0 \qquad \sum F_y = 0 \qquad \sum M_z = 0$$

Concept Reinforcement

1. List and discuss the requirements for equilibrium

2. Explain how angles are defined in force systems

3. List and describe the four types of equilibrium problems

Section 1.13 – Three-Dimensional Equilibrium

Section Objective

- Explain how to solve three-dimensional equilibrium problems

Introduction

This section presents the concept of a three dimensional force system in equilibrium. For the system to be in equilibrium, all the forces in all directions (3-dimensional space) must sum to zero and the moments associated with these forces must also sum to zero.

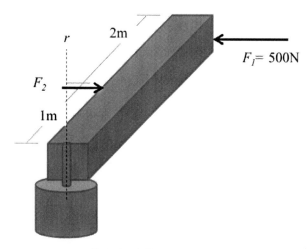

Original force system

Equilibrium

Mechanical equilibrium is a condition of an object where the sum of the forces acting on the object is zero and the sum of the moments acting on the object is zero.

Mechanical Equilibrium: $\sum F = 0$ and $\sum M = 0$

When considering three-dimensional equilibrium we have a more complicated set of requirements for equilibrium to exist:

Mechanical Equilibrium of forces: $\sum F_x = 0$, $\sum F_y = 0$ and $\sum F_z = 0$.

Mechanical Equilibrium of moments: $\sum M_x = 0$, $\sum M_y = 0$ and $\sum M_z = 0$.

The vector sum of the external forces and force components is zero in the x-direction, in the y-direction and in the z-direction. Forces in a three dimensional (x, y, z) system act in any conceivable direction but can be deconstructed and presented as x, y and z components of the force.

The sum of the moments acting on the object is zero in the x-direction, in the y-direction and in the z-direction. Moments in a three dimensional (x, y, z) system act in any conceivable direction but can be deconstructed and presented as x, y and z components of the moment.

Because of the assumption that the object is a rigid body this definition can be expanded.

Mechanical equilibrium is a condition of an object where the vector sum of the forces acting on all the particles of the object is zero and the sum of the moments acting on all the particles of the object is zero.

Newton's Laws of Motion

Isaac Newton developed several laws of motion that relate to the issue of forces and equilibrium.

Newton's First Law of Motion:

A particle that is at rest remains at rest unless acted upon by an outside force.

A particle that is moving with a constant velocity continues to move with constant velocity unless acted upon by an outside force.

These statements both assume that the outside force is a single force not countered by an equal and opposite force.

The first law supports the concept that an object in equilibrium must have zero resultant force and zero moment acting on it.

Newton's Second Law of Motion:

The force (F) applied to an object of mass (m) causes acceleration (a) of the object in the direction of the force vector and the relationship is: $F = ma$.

The second law explains that if a force is applied the object must be accelerated.

Newton's Third Law of Motion:

For every action force of object X on object Y there is a reaction force of object Y on object X, which is equal in magnitude, opposite in direction and acts at exactly the same time.

The third law explains that to have an object in mechanical equilibrium requires equal and opposite forces between the object and its environment.

Zero point for Angles

In order to maintain a consistent treatment for the direction of forces and moments, the angles will be defined from the horizontal x-axis pointing to the right. This is established as zero degrees (0°) and the magnitude of the angle increases in the counterclockwise direction until the angle of the vector is pointing again directly to the right. At this point the vector has traversed one full rotation and is at an angle of 360° which is equivalent to 0°.

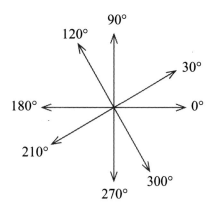

In a 3-dimensional Cartesian coordinate system, the angle between vectors is what is important when evaluating force and moment components.

Three-Dimensional Equilibrium

This section focuses on equilibrium in a three-dimensional system in the x-y-z Cartesian coordinate system. Forces can act in the x, y or z directions or any angular direction in the coordinate system. Therefore moments will act perpendicular to the associated force but this direction can also be any conceivable direction. In a system in equilibrium, all the forces and force components sum to zero and all the moments sum to zero.

To solve these equilibrium problems we need to identify the forces and force components acting in the x-direction, in the y-direction and in the z-direction. The objective is to determine the amount of force in each direction which will result in mechanical equilibrium where the sum of the forces and the moments is zero:

Mechanical Equilibrium of forces: $\sum F_x = 0$, $\sum F_y = 0$ and $\sum F_z = 0$.

Mechanical Equilibrium of moments: $\sum M_x = 0$, $\sum M_y = 0$ and $\sum M_z = 0$.

Three-Dimensional Equilibrium problems

Example 1: 3D Equilibrium

In the original force system, there is a force (F_1) of 500 N acting perpendicular on an arm of length 3 m. There is a separate force (F_2) acting perpendicular on the arm at a length of 1 m but from the opposite direction. If the two forces are replaced with a force and a couple at the axis (r) of the arm and the moment at the axis point is 900 Nm, calculate the force (F_2), and the force (F_A) acting on the axis point.

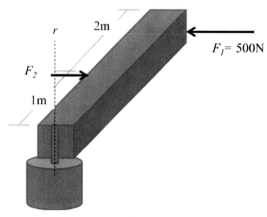

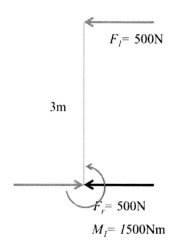

Original force system

First, determine the force and couple system associated with the force F_1 once it is moved to the radial position (r).

The force shifted to the axis point r will be a 500 N force in the same direction as F_1.

Next, determine the couple and the magnitude of the moment associated with the couple that will be acting at r.

Force F_1 replaced with a force and couple at r

The couple is a pair of 500 N forces each at 1.5 m from the center point. Together these add:

$M = Fr = 500$ N (3 m) = 1,500 Nm

The resultant moment at the point r is given as 900 Nm. Therefore we must find a force (F_2) that, when shifted to the point r, gives a moment in the opposite direction of the difference: $M_R = M_1 - M_2$

Solving for M_2 we have $M_2 = M_1 - M_R = 1,500$ Nm $- 900$ Nm $= 600$ Nm

The formula for the moment from F_2 is: $M = Fr = F_2 (1\text{m})$

With $M_2 = 600$ Nm and solving for F_2 we have: $F_2 = M/(1$ m$) = 600$ Nm/1 m $= 600$ N.

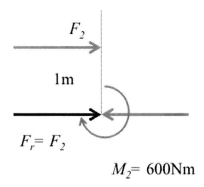

$$M_2 = 600\text{Nm}$$

Force F_2 replaced with a force and couple at r

Therefore, the force (F_2) is equal to 600 N in the opposite direction.

The final step is to determine the force acting on the point r.

We have F_1 (500 N) acting from the right and F_2 (600 N) acting from the left.

$F_r = F_1 + F_2$ but they are from opposite directions and therefore:

F_r = 500 N + 600 N = 100 N acting from the left.

The final force system has a force of 600 N acting at point r from the left and a moment of 900 Nm acting in a counterclockwise direction on point r.

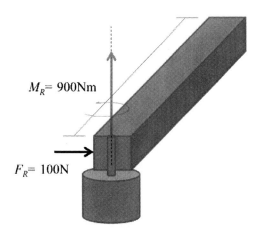

Final force system

Types of Three-Dimensional Equilibrium Problems

There are four general types of three-dimensional equilibrium problems. These are the concurrent at a point collinear force system, the parallel force system, the concurrent force system and the general force system.

Force System Concurrent at a Point

In a force system that is concurrent at a point all the forces are acting from the exact same origin. The illustration presents how this might appear. With a multitude of forces all acting from the same origin, the forces must sum to zero for equilibrium to exist.

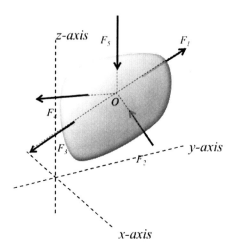

Point concurrent force system

$$\sum F_x = 0, \sum F_y = 0, \sum F_z = 0$$

This system, because all the forces are acting from the same point, has no moments.

Force System Concurrent at a Line

In a force system that is concurrent at a line all the forces are acting on the exact same axis in the object. The illustration presents how this might appear. With a multitude of forces all acting along the same line of action the forces must sum to zero for equilibrium to exist.

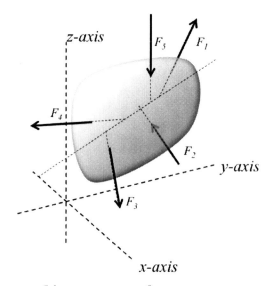

Line concurrent force system

$$\sum F_x = 0, \sum F_y = 0, \sum F_z = 0$$
$$\sum M_x = 0, \sum M_y = 0, \sum M_z = 0$$

This force system will have moments.

When the axis line is along one of the major axes, for example the x-axis this reduces to

$$\sum F_x = 0, \sum F_y = 0, \sum F_z = 0$$
$$\sum M_y = 0, \sum M_z = 0$$

Parallel Force System

In a parallel force system all the forces act on parallel lines of action. The illustration presents how this might appear. With a multitude of forces all acting along parallel lines of action the forces must sum to zero and the moments must sum to zero for equilibrium to exist.

This force system will have moments.

When the axis line is along one of the major axes, for example the x-axis the equilibrium requirements reduce to

$$\sum F_x = 0$$
$$\sum M_y = 0, \sum M_z = 0$$

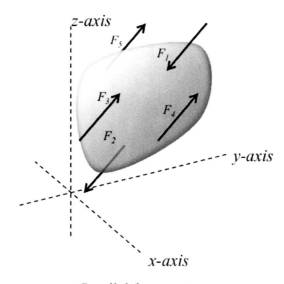

Parallel force system

$$\sum F_x = 0, \sum F_y = 0, \sum F_z = 0$$
$$\sum M_x = 0, \sum M_y = 0, \sum M_z = 0$$

In a general force system all the forces are acting in random directions on the object and various moments are produced. The illustration presents how this might appear. With a multitude of forces all acting along different lines of action the vector sum of the forces and force components must equal zero for equilibrium to exist. Also, in this case the moments about the *x, y* and *z*-axes must sum to zero for equilibrium to exist.

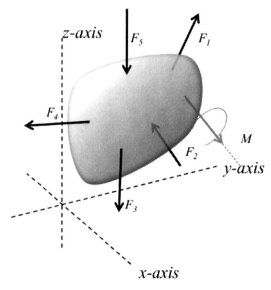

General force system

$$\sum F_x = 0, \sum F_y = 0, \sum F_z = 0$$

$$\sum M_x = 0, \sum M_y = 0, \sum M_z = 0$$

Concept Reinforcement

1. Explain the requirements for equilibrium to exist in a 3-dimensional force system

2. Explain how angles are measured in a 3-dimensional Cartesian coordinate system.

3. Explain the four types of 3-dimensional force systems.

Section 1.14 – Trusses

Section Objective

- Explain trusses and their uses

Introduction

This section presents the concept of the truss, a structural element that is utilized in many different building and construction applications. Trusses use triangular configurations of beams because the triangle is a more stable shape than a square or rectangle. All trusses experience both compression and tension in different structural elements. There are two general styles of trusses: triangular and flat.

Truss

A truss is a support structure used to create mechanical stability. The most basic form of a truss is three support beams connected in a triangular shape. Under moderate levels of stress, a triangle will not change shape as easily as other simple geometric shapes like a square. Therefore, the forces that act on a truss will produce strain and moments in the beams, but the overall structure will not be seriously affected.

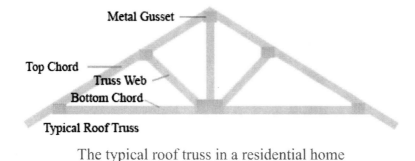

The typical roof truss in a residential home

Forces Acting on a Truss

Within all trusses and all structures, the forces that are experienced can be segregated as that of tension (pulling) and compression (pressing). The structural materials that are better able to withstand tension are used in places in the truss where tension is the predominant force experienced. Those materials that are better at handing compression are therefore positioned where the primary force experienced is compression.

Tension

Tension is a pulling force where the two ends of a beam are drawn apart. With this type of force the structural member that is experiencing this stress is in "tension." The properties of the material that makes up the structural member dictate how well the beam can withstand these forces of tension. This is also dependent upon the size, shape and cross-sectional area of the beam. Some materials are better able to withstand tension and some are better with compression.

Tension is the force typically experienced by the structural beams at the underside of a truss when a load is experienced by the truss from above.

A structural beam experiencing tension

Compression

Compression is a pressing force where the two ends of a beam are forced towards each other. With this type of force the structural member that is experiencing this stress is in "compression." The properties of the material that makes up the structural member dictate how well the beam can withstand these forces of compression. This is also dependent upon the general size, length, shape and cross-sectional area of the beam. Some materials are better able to withstand compression and some are better with tension.

A structural beam experiencing compression

Compression is the force typically experienced by the beams on the top of a roof or in the support columns in a building.

Two General types of Truss

There are two major families of truss shapes, which are described by the outer shape of the overall structure rather than the inner webbing within the structure. The triangular truss is a very common shape in residential homes and many other applications. The flat shape truss is typical of the structure used in floors and flat construction.

Within all trusses, the "webbing" is built as triangular units for the same reasons of the stability of the triangle structure.

Triangular Shape

As mentioned above, the outer structure is what segregates these general truss types. The triangular structure is useful in residential structures, primarily for roof construction. The top side of the truss bears the load of rain, snow and wind plus the weight of the structure itself. Therefore, the top beams are in compression as the load presses on the beams. The smaller webbing beams that rise up and connect to the middle of the top members are also experiencing the load and compression.

While the load presses down from above, the bottom beams are being pulled from both ends, experiencing tension. This is the same for the two longer webbing sections that rise to meet at the peak. They are stretched and therefore experience tension as well.

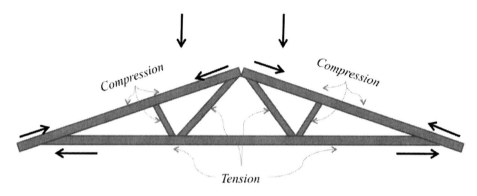

A common truss with the forces of compression and tension presented

There are many variations on this structure. The variants all have different names. Generally, the styles differ by the way the internal webbing beams are structured. When the size expands to the point that more than a single triangular webbing is required, the styles are then called double and triple trusses.

Some common names are the King Post, Queen Post, the Fan, the Fink, and the Howe. Larger roofs use expanded versions of these same styles with names such as Double Fan, Modified Queen, Double Fink, Double Howe, Triple Fink, Triple Howe, etc.

Flat Shape

The flat truss is made of two parallel beams separated by consistently shaped triangular sections of cross beams, or webbing. This is the form commonly visible in flooring, but also commonly used with bridges, particularly with the railroads. One common type of flat truss is the Pratt Truss.

In the Pratt Truss the vertical sections are there to absorb the forces of compression and the non-vertical sections are there to absorb the forces of tension.

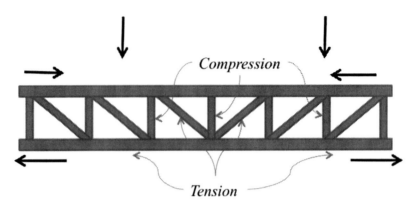

A Pratt truss with the forces of compression and tension presented

There are also several different styles of flat trusses with some using a cantilever (an overhang) some using only angular webbing (no vertical posts) and some using non-parallel horizontal beams.

When this form is applied to bridges, there are several names that are common; Pratt truss, the Curved Chord Pratt truss, the Baltimore truss, the Pennsylvania (Pratt) truss, the Warren truss, the Quadrangular Warren truss, the Subdivided Warren Type truss, the Lattice truss and the Whipple truss.

Concept Reinforcement

1. Describe a truss.

2. Explain the two general types of truss.

3. Explain why a triangular structure is most common.

4. Explain where the forces of compression and the forces of tension are most prevalent in a basic triangular roof truss.

Section 1.15 – Applications

Section Objective

- List applications of mechanical engineering

Introduction

This section presents applications of force vectors and moments related to force couples and force systems.

Vector Applications

Multiplying Force Vectors

A force (F_1) of 200 N is acting horizontally east on an object. The force is to be multiplied by 5. Calculate the new force and the direction of the force vector.

Solution:

Multiplying a vector by a scalar only changes the magnitude of the vector while the direction remains the same.

New force = $5F_1$ = 5(200 N) horizontally east = 1,000 N horizontally east.

Adding Force Vectors

With vector1 having a magnitude 5 with a direction of east and vector2 having a magnitude of 8.66 with a direction of north, calculate the resultant vector.

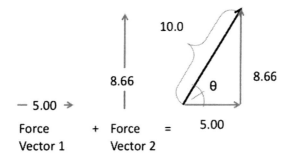

Solution:

We draw the first vector. Then, starting where the first vector ends, we draw the second vector. In this case the second vector is in a direction that is 90° from the first vector.

This requires the use of trigonometry and the Pythagorean Theorem for right triangles, where $A^2 + B^2 = C^2$. Vector 1 is A, Vector 2 is B and the resultant vector is C. Solving for C (the magnitude of the resultant vector):

$$C = \sqrt{A^2 + B^2} = \sqrt{5.00^2 + 8.66^2} = \sqrt{100} = 10.0$$

The angle can be determined through various trigonometry functions:

Arc tangent: arctan (8.66/5.00) = 60°

Arc sine: arcsin (8.66/10.0) = 60°

Arc cosine: arccos (5/10.0) = 60°

The resultant vector is 10.0 ∡ 60° NE.

Free-Body Diagram

An object is on a ramp with an incline. It has a force due to the acceleration of gravity = 5,000 N, there is a normal force which opposes the fraction of the force of gravity against the ramp, an external force acting on the object down the ramp which is a fraction of the force of gravity and a force of friction of 300 N which is acting opposite the direction of movement. The ramp is at a 30° angle. Define the magnitude of the other forces.

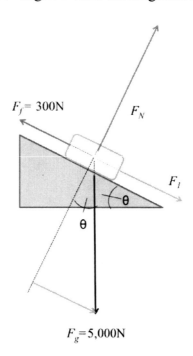

Free-body diagram of an object on a ramp

Solution:

First, define the formula for the force of gravity pulling the object down the ramp.

As can be seen in the figure, the force F_1 that is pulling the object down the ramp is related to the force of gravity.

With the force of gravity being the hypotenuse of the right triangle below the ramp we have: $F_1 = F_g \sin\theta$

This equals: $F_1 = F_g \sin\theta = 5$ kN $(\sin 30°) = 5$ kN $(0.5) = 2.5$ kN

The normal force is also a fraction of the force of gravity.

With the force of gravity being the hypotenuse of the right triangle below the ramp we have: $F_N = F_g \cos\theta$

This equals: $F_N = F_g \cos\theta = 5$ kN $(\cos 30°) = 5$ kN $(0.866) = 4.3$ kN

Moving a Force off its line of action

In this exercise, we replace a force acting at a point with a parallel force and a couple at a different point on the object. The force $F_1 = 300$ N. The distance between the points is 20 cm. The angle $\theta = 15°$. Calculate the equivalent force system in which a parallel force is moved to point B and a moment acts at point B.

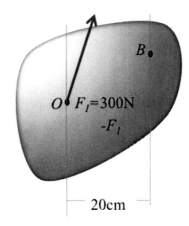

Original Free-body diagram of an object with a force F_1

Solution:

First, we move the force F_1. See the illustration where the force F_1 has been moved to point B and an opposing force $-F_1$ has been placed at point B. This is an equivalent system to the original system

Next we consider how to calculate the magnitude of the moment that acts at point B.

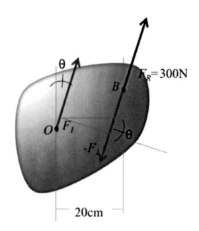

An equivalent system where the force F_1 is being moved to point B

Because we are using the horizontal measure of 20 cm as the distance between the two points we can determine the vertical component of the force vector (the "up" direction in the illustration). The other option is to determine the separation distance that is perpendicular to the force vectors. Either way we use the cosine of the angle θ of 15°.

$M = (F\cos\theta)r$ or $M = Fr\cos\theta$

The moment is therefore:

$M = (F\cos\theta)r = 300 \text{ N} (\cos 15)0.20 \text{ m} = 300 \text{ N} (0.966)0.20 \text{ m} = 57.9 \text{ Nm}$

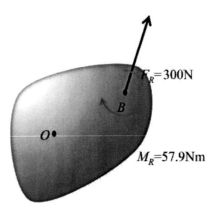

An equivalent system where the force F_R is moved to point B
and the Moment M_R acts on point B

Concept Reinforcement

1. Calculate the result of multiplying the acceleration vector with a magnitude of 9.8 m/s² in the direction towards earth, by the mass scalar with a magnitude of 100 kg.

2. Add two force vectors: a vector with a magnitude of 173 and a direction directly east and a vector with a magnitude of 100 and a direction directly north.

3. Create a free-body diagram: An object is on a ramp with an incline. It has a force due to the acceleration of gravity = 2,500 N, there is a normal force which opposes the fraction of the force of gravity against the ramp, an external force acting on the object down the ramp which is a fraction of the force of gravity and a force of friction of 800 N which is acting opposite the direction of movement. The ramp is at a 60° angle. Create the free body diagram and calculate the magnitude of the normal force and the force pulling the object down the ramp.

4. Replace a force acting at a point with a parallel force and a couple at a different point on the object. The force F_1 = 4,500 N. The distance between the points is 35 cm. The force vector acts perpendicular to the measurement of the separation distance. Calculate the equivalent force system in which a parallel force is moved from the original point O to point B and a moment acts at point B.

Unit Two

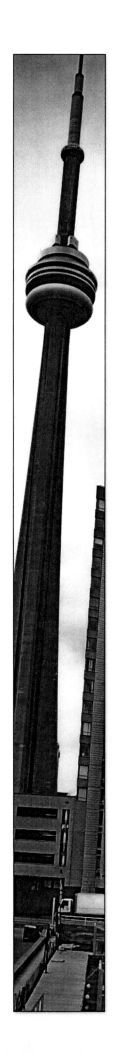

Section 2.1 – Force Analysis in Joints   104

Section 2.2 – Section Analysis   111

Section 2.3 – Frames and Machines   117

Section 2.4 – Centroids of Areas   127

Section 2.5 – Centroids of Composite Areas   133

Section 2.6 – Distributed Loads   139

Section 2.7 – Centroids of Volumes and Lines   143

Section 2.8 – Pappus-Guldinus Theorems   149

Section 2.9 – Center of Mass, Simple Objects   153

Section 2.10 – Center of Mass of Composites   159

Section 2.11 – Moment of Inertia   165

Section 2.12 – Moments of Simple Objects   169

Section 2.13 – Rotated and Principal Axes   173

Section 2.14 – Parallel Axis Theorem   177

Section 2.15 – Analysis Application   181

Section 2.1 – Force Analysis in Joints

Section Objective

- Explain the method of joints

Introduction

This section presents the method of joints, one of two methods to analyze and resolve the various forces and moments that are acting on a truss. The method of joints assumes all forces act at the joints only. The method of joints initially resolves the external forces acting on the truss using the standard equilibrium equations where the forces in each direction sum to zero and the moments about any point sum to zero. After this, the forces acting at each joint are resolved one at a time until all forces are identified. The end result is a free-body diagram with all the forces displayed acting at each joint.

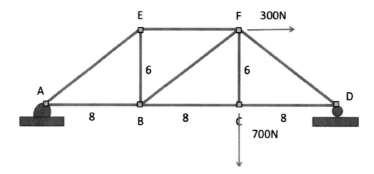

Ideal Truss

For the purposes of simplifying mechanical force analysis of trusses, a concept called an ideal truss is defined.

In an ideal truss:

1. all sections are weightless; i.e. the weight of the sections do not impact the equilibrium calculations

2. all sections lie in a two dimensional plane; i.e. this is only a two dimensional problem

3. all sections are connected by smooth pins; i.e. the connections themselves do not change the forces or moments

4. all sections form triangular substructures within the main structure

5. all external loads are applied only at the joints

The Rigid Body Assumption is also held. This states that the sections and the entire truss acts as a rigid body (that it does not twist or bend).

Mechanical Analysis of Forces

To analyze the forces acting on an object the standard procedure is to create a free-body diagram with all the forces acting from the geometric center of the object. Ideally all the moments would also be translated to the geometric center of the object as well.

For the mechanical analysis of the forces acting on a truss there are two primary procedures: 1. The Method of Joints and 2. The Method of Sections.

The Method of Joints applies the concepts of an Ideal Truss, assumes all forces act at a joint (a connection of two or more sections), and assumes all moments are the result of forces acting at joints. The objective of the Method of Joints is to define the forces and moments acting at the joints in question. This assumes that the forces are in balance and that the truss is in mechanical equilibrium; i.e. all forces in all directions sum to zero and all moments in all directions also sum to zero.

The Method of Sections also applies the concepts of an Ideal Truss, assumes all forces act at a joint (a connection of two or more sections), and assumes all moments are the result of forces acting at joints. The objective of the Method of Sections is to define the force's action on the various sections of the Truss. Whether the section is in compression or is in tension is determined through the Method of Sections.

We will focus on the Method of Joints.

Mechanical Equilibrium

Mechanical equilibrium is a condition of an object where the sum of the forces acting on the object and the sum of the moments acting on the $\sum F = 0$ and equation $\sum M = 0$.

When considering two-dimensional equilibrium, we have:

Mechanical Equilibrium: equation $\sum F_x = 0$, $\sum F_y = 0$ and equation $\sum M_z = 0$.

Where:
F_x = a force or force component acting the x-direction in newtons (N)
F_y = a force or force component acting the y-direction in newtons (N)
M_z = a moment due to the forces acting in the x-y plane and acting the z-direction in newton-meters (Nm)

The vector sum of the external forces and force components in the x-direction is zero and the sum of the external forces and force components in the y-direction is zero. The sum of the moments acting on the object in the z-direction is zero. All moments in a two-dimensional (x, y) system act in the z-direction.

Zero point for Angles

In order to maintain a consistent treatment for forces and moments, the angles will be defined from the horizontal vector pointing to the right. This is established as zero degrees (0°) and the magnitude of the angle increases in the counterclockwise direction until the angle of the vector is pointing again directly to the right. At this point the vector has traversed one full rotation and is at an angle of 360° which is equivalent to 0°.

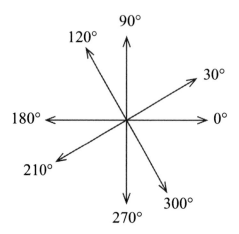

Method of Joints

Within the analysis the structure of the truss is assumed to be a rigid body (i.e. it does not deform), the connecting joints are all smooth frictionless pins (i.e. they do not add or detract from the external force or moment) and the individual sections are weightless (i.e. the weight is only externally applied).

Regarding the base of support, there are generally two types of structure support, a pinned joint and one that is resting on a roller. The pinned join can experience forces in both the x and the y directions (either up or down). Supports that rest on a roller can only experience compression in the single direction, usually vertically down. Because the roller allows horizontal movement these joints do not experience horizontal (x-direction) forces and do not experience moments.

The Method of Joints is best expressed through an example.

Example 1: Method of joints

With a truss as presented in the figure with horizontal members of 8 m and vertical members of 6 m, there is a force applied horizontally at point F of 300 N and a force applied vertically at point C of 700 N.

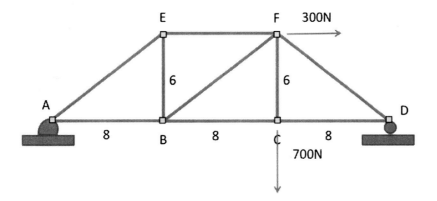

The first step is to create a free-body diagram of the structure. This is presented in the second figure.

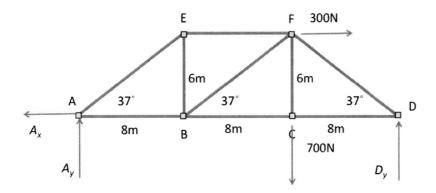

The second step is to resolve all the forces into their x and y components. This is already done in the free-body diagram. Note that the support on the left has both an x and a y component because it was a pinned joint while the support at the right was only a roller and therefore only has a vertical component to the force.

The third step is to write and solve the three equilibrium equations: equation $\sum F_x = 0$, equation $\sum F_y = 0$ and equation $\sum M_z = 0$.

Forces in the x-direction:

We define that force acting to the right is positive and force acting to the left is negative.

The free-body diagram shows 300 N positive and a force component of A_x being negative.

$$\sum F_x = -A_x = 300 \text{ N}$$

Because the total force at equilibrium is zero, we therefore conclude that $A_x = 300$ N

Forces in the y-direction:

We define that force acting in the "up" direction is positive and force acting in the "down" direction is negative.

The free-body diagram shows positive forces of A_y and D_y and a negative force of 700 N.

$$\sum F_y = A_y + D_y - 700 \text{ N}$$

Because the total force at equilibrium is zero, we therefore conclude that $A_y + D_y = 700$ N

Moments in the z-direction:

We define that a moment acting in the "clockwise" direction is positive and a moment acting in the "counterclockwise" direction is negative.

The free-body diagram shows positive moments due to the force D_y acting at a distance of 24 m from point A, a negative moment due to the force of 700 N at point C a distance of 16 m from point A and a second negative moment due to the force of 300 N at point F a distance of 6 m vertically from point A.

$$\sum M_z = D_y (24 \text{ m}) - 700 \text{ N} (16 \text{ m}) - 300 \text{ N} (6 \text{ m})$$

Because the total moment at equilibrium is zero we therefore conclude that

$$D_y = \frac{700 \text{ N} (16 \text{ m}) - 300 \text{ N} (6 \text{ m})}{24 \text{ m}} = 542 \text{ N}$$

And from $A_y + D_y = 700$ N we have $A_y = 158$ N.

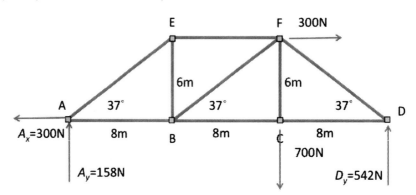

Free-Body Diagram with all external forces defined

The Method of Joints now comes into play to determine the forces acting in the internal members. This is a type of analysis that focuses on each joint (pinned or on a roller) separately. There are no moments acting at a joint, therefore the solution is resolved through the force equilibrium equations.

$$\sum F_x = 0, \sum F_y = 0$$

With only two equations, we can solve for only two unknown forces at any specific joint. The only joints that have only two forces acting are point A and point D, so this is where we start.

Starting with Point A, we again use the process of creating a Free-Body Diagram, then resolving for the forces and solving the equilibrium equations.

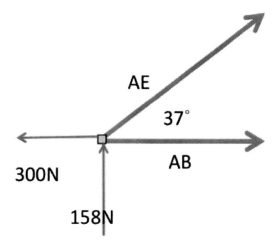

Free-Body Diagram of Point A

Here is the Free-Body Diagram for Point A and the FBD with the forces resolved into their *x*- and *y*-components.

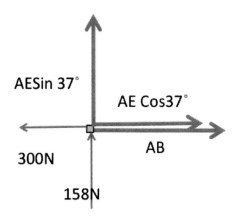

Free-Body Diagram of Point A with forces resolved

Solving for the force equilibrium we have $\sum F_x = 0, \sum F_y = 0$

$\sum F_x = AE\cos 37° + AB - 300\text{ N} = 0$ and $\sum F_y = AE\sin 37° + 158\text{ N} = 0$

Solving for *AE* we have $AE = -\dfrac{158\text{ N}}{\sin 37°} = -263\text{ N}$

Here the negative direction means that the member *AE* is in tension, the force is directed towards the point A.

Solving for *AB* we have $AB = 300\text{ N} - AE\cos 37° = 300\text{ N} - (-210\text{ N}) = 510\text{ N}$

Therefore, the force is also in tension. This time, however, because of convention (we have positive to the right) the magnitude is positive.

Using this procedure, all the forces acting at all the joints can be resolved.

Concept Reinforcement

1. Explain the assumptions for an Ideal Truss.

2. Explain Mechanical Analysis of Forces.

3. Explain Mechanical Equilibrium.

4. Describe the convention for angles in mechanical systems.

Section 2.2 – Section Analysis

Section Objective

- Explain the method of sections

Introduction

This section presents the method of sections, one of two methods to analyze and resolve the various forces and moments that are acting on a truss. The method of sections assumes all forces act at the joints only. The method of sections initially resolves the external forces acting on the truss using the standard equilibrium equations where the forces in each direction sum to zero and the moments about any point sum to zero. After this, a cut across the truss at the location of interest is made and the forces acting on those members at each joint are resolved along with the moments. The end result is a free-body diagram with the forces displayed acting on the members (sections) of interest.

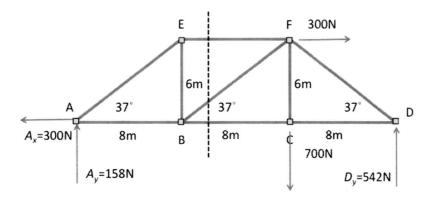

Free-Body Diagram with all external forces defined
and the truss "cut" across members EF, BF, and BC

Mechanical Analysis of Forces

To analyze the forces acting on an object the standard procedure is to create a free-body diagram with all the forces acting from the geometric center of the object. Ideally, all the moments would also be translated to the geometric center of the object as well.

For the mechanical analysis of the forces acting on a truss there are two primary procedures; 1. The Method of Joints and 2. The Method of Sections.

The Method of Joints applies the concepts of an Ideal Truss, assumes all forces act at a joint (a connection of two or more sections), and assumes all moments are the result of forces acting at joints. The objective of the Method of Joints is to define the forces and moments acting at the joints in question. This assumes that the forces are in balance and that the truss is in mechanical equilibrium; i.e. all forces in all directions sum to zero and all moments in all directions also sum to zero.

The Method of Sections also applies the concepts of an Ideal Truss, assumes all forces act at a joint (a connection of two or more sections), and assumes all moments are the result of forces acting at joints. The objective of the Method of Sections is to define the force's action on the various sections of the Truss. Whether the section in compression or is it in tension is determined through the Method of Sections.

We will now focus on the Method of Sections.

Mechanical Equilibrium

Mechanical equilibrium is a condition of an object where the sum of the forces acting on the object and the sum of the moments acting on the object are both zero. In summary this appears as mechanical equilibrium: $\sum F = 0$ and $\sum M = 0$ and when considering two-dimensional equilibrium we have:

Mechanical Equilibrium: $\sum F_x = 0$, $\sum F_y = 0$ and $\sum M_z = 0$.

Where:
F_x = a force or force component acting in the x-direction in newtons (N)
F_y = a force or force component acting in the y-direction in newtons (N)
M_z = a moment due to the forces acting in the x-y plane and acting the z-direction in newton-meters (Nm)

The vector sum of the external forces and force components in the x-direction is zero and the sum of the external forces and force components in the y-direction is zero. The sum of the moments acting on the object in the z-direction is zero. All moments in a two-dimensional (x, y) system act in the z-direction.

Method of Sections

In the method of sections a line is draw through the truss across several members. This line can be a straight line or a curved line. The number of members "cut" can be many. After the truss has been "divided" a free body diagram is made, the x and y components of the forces are resolved and the equilibrium equations solved.

Example 1: Method of Sections

With a truss, as presented in the figure, with horizontal members of 8 m and vertical members of 6 m, a force is applied horizontally at point F of 300 N and a force applied vertically at point C of 700 N.

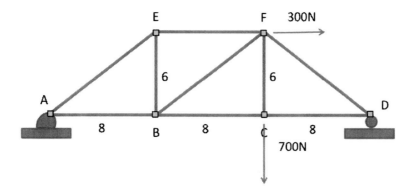

Initially, before the method of section can be applied, the external forces acting on the truss must be solved. The first step is to create a free-body diagram of the structure. This is presented in the second figure.

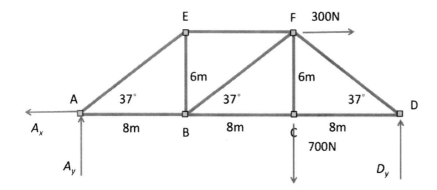

The second step is to resolve all the forces into their x and y components. This is already done in the free-body diagram. Note that the support on the left has both an x and a y component because it was a pinned joint, while the support at the right was only a roller and therefore only has a vertical component to the force.

The third step is to write and solve the three equilibrium equations: $\sum F_x = 0$, $\sum F_y = 0$ and $\sum M_z = 0$.

Forces in the x-direction:

We define that force acting to the right is positive and force acting to the left is negative.

The free-body diagram shows 300 N positive and a force component of A_x being negative.

$$\sum F_x = -A_x + 300 \text{ N}$$

Because the total force at equilibrium is zero, we therefore conclude that $A_x = 300$ N.

Forces in the y-direction:

We define that force acting in the "up" direction is positive and force acting in the "down" direction is negative.

The free-body diagram shows positive forces of A_y and D_y and a negative force of 700 N.

$$\sum F_x = A_y + D_y - 700 \text{ N}.$$

Because the total force at equilibrium is zero, we therefore conclude that $A_y + D_y = 700$ N.

Moments in the z-direction:

We define that a moment acting in the "clockwise" direction is positive and a moment acting in the "counterclockwise" direction is negative.

The free-body diagram shows positive moments, which are due to the force D_y, acting at a distance of 24 m from point A, a negative moment due to the force of 700 N at point C, a distance of 16 m from point A, and a second negative moment due to the force of 300 N at point F, a distance of 6 m vertically from point A.

$$\sum M_z = D_y (24 \text{ m}) - 700 \text{ N} (16 \text{ m}) - 300 \text{ N} (6 \text{ m})$$

Because the total moment at equilibrium is zero we therefore conclude that

$$D_y = \frac{700 \text{ N} (16 \text{ m}) - 300 \text{ N} (6 \text{ m})}{24 \text{ m}} = 542 \text{ N}$$

From $A_y + D_y = 700$ N we have $A_y = 158$ N.

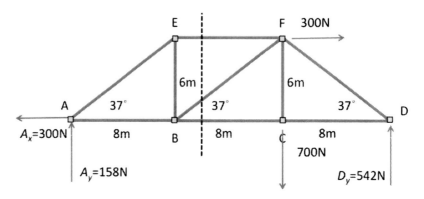

Free-Body Diagram with all external forces defined
and the truss "cut" across members EF, BF, and BC

The method of section can now be applied using the following steps.

1. Produce a free-body diagram.

2. Resolve the x and y components of the forces.

3. Solve the force equilibrium equations.

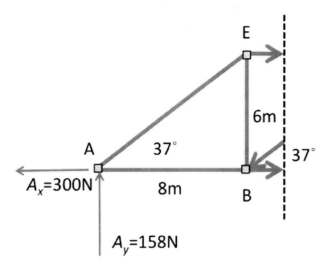

Free-Body Diagram with all external forces defined
and the internal forces of the truss in members EF, BF, and BC resolved

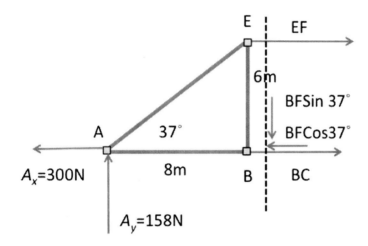

Free-Body Diagram with all external forces defined
and the truss "cut" across members EF, BF, and BC

Solving for the force equilibrium, we have $\sum F_x = 0$, $\sum F_y = 0$ and $\sum M_z = 0$.

$\sum F_x$ = -BF cos37° + BC − 300 N = 0 and $\sum F_y$ = -BF cos37° + 158 N = 0 and $\sum M_z$ = 158 N (8 m) + EF (6 m) = 0.

Solving for *EF*, we have: $EF = \dfrac{-158 \text{ N (8 m)}}{6 \text{ m}} = -211 \text{ N}$

Here, the negative direction means that the member *EF* is in tension, the force is directed towards the point E.

Solving for *BF* we have $BF = -\left(-\dfrac{158 \text{ N}}{\sin 37°} \right) = 263 \text{ N}$

Here the positive direction means that the member *BF* is in the direction we specified (downward) and therefore is in tension, the force is directed towards the point B.

Solving for *BC* we have $BC = 300 \text{ N} - BF \cos 37° = 300 \text{ N} - (-210 \text{ N}) = 510 \text{ N}$.

Therefore the force is also in tension but this time, because we initially selected the direction of the force to the left (we have positive to the right) the magnitude is positive.

This procedure allows one to resolve all the internal forces acting in the sections.

Concept Reinforcement

1. Explain the types of mechanical analysis of trusses.

2. Explain the method of sections.

3. Explain the difference between the method of joints and the method of sections.

Section 2.3 – Frames and Machines

Section Objective

- Explain frames and machines

Introduction

This section presents the concepts of frames and machines. Frames are structures similar to trusses. Both are expected to be static (stationary) but the pattern of the structure of frames is more open and flexible than trusses. Frames can carry loads as well as trusses. This section also presents the machine, which is a general term for a mechanical structure that includes moving parts and is not expected to be static.

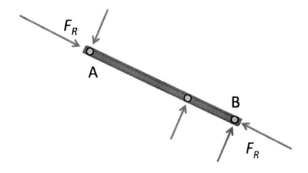

Non-Axial member (resultant forces)

Definition of Statics

Within mechanical engineering, statics focuses on the analysis of external forces and moments on structures that are in static equilibrium. In a structure (a physical system) in static equilibrium, all the sections, components, members, etc. do not move relative to each other over time. Either the entire system does not move over time or all of it moves at the same velocity over time. Either way, the relative positions do not change and the way the subsections experience the external forces does not change over time.

Mechanical Analysis

Applying mechanical analysis of a structure to problems of statics initially involves several key assumptions:

1. the structure and the substructures are assumed to act as a rigid body (i.e. they do not deform),

2. the connecting joints are all smooth frictionless pins (i.e. they do not add or detract from the externally applied forces or moments)

3. the individual sections are weightless (i.e. the loads are only externally applied at the specific points of application of the external forces).

These assumptions are used in this early mechanical analysis so that major mechanical principles can be the focus of the analysis. In more advanced mechanical analysis, these assumptions will be eliminated one by one.

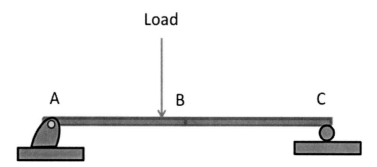

Structure pinned at the left support, with a roller as the right support

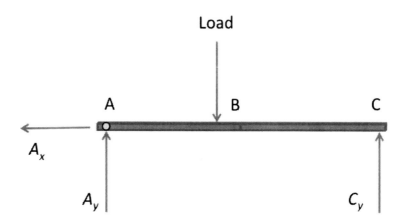

Free Body Diagram of the same structure

Structures rely on various types of support, such as attachment mechanisms to the wall, floor or ceiling. There are two general types of structural support: a pinned joint and a joint that is resting on a roller. The pinned joint can experience forces in both the x and the y directions and be in either the positive or negative direction. Supports that rest on a roller can only experience compression in the single direction. The compression is directed into the support. Because the roller allows movement in the other direction, these joints experience forces in one direction only and do not experience moments.

Axial vs Non-Axial Member

The structural members of a mechanical system experience forces in two primary ways: only at the ends or in more than just at the ends.

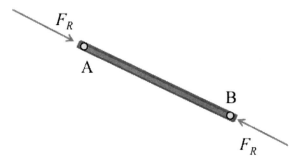

Axial member (resultant forces)

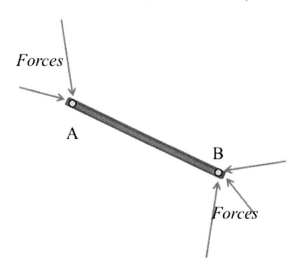

Axial member (various forces)

An Axial Member is one that experiences the forces only at the ends and the forces can be summed so that the line-of-action is along the member, which will show that the member is only in tension or in compression.

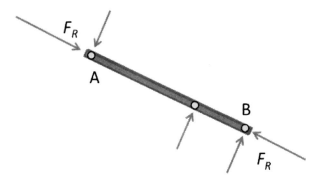

Non-Axial member (resultant forces)

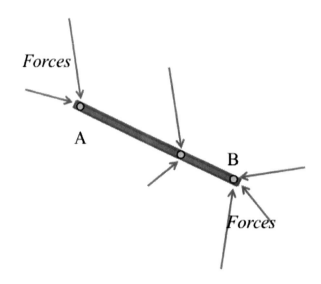

Non-Axial member (various forces)

A non-axial member is one that also experiences external forces in the center span of the member and hence the line-of-action of the forces are not only along the length of the member. In this case, the resultant forces (the sum of the forces) are shown to have a line-of-action along the member and a line-of-action that is perpendicular to the member.

Frames

A frame is a set of structural members and is usually considered separate from a truss. A truss is a structure that includes substructures within the outer framework of the truss itself.

In the analysis of the forces in a frame that is in mechanical equilibrium there are three typical steps;

1. Create a Free-Body Diagram

2. Identify all the forces and identify the x and y components of each force

3. Solve the three equilibrium equations:

   A. $\sum F_x = 0$
   B. $\sum F_y = 0$
   C. $\sum M_z = 0$

Simple Machines

There are six simple machines, which are all methods to improve the mechanical output of a mechanical input. These machines are inclined planes, the three types of levers, the pulley, the wedge, the wheel and axle, and the screw.

Inclined Plane

An inclined plane is designed for two purposes. One is to reduce the external work required to move an object horizontally by utilizing gravity to reduce the apparent weight and the frictional forces resisting movement. In the graphic, the direction of the movement is down the incline. As the angle of inclination increases, the amount of external work required decreases. (when $\theta = 0°$, $\cos\theta = 1$ and when $\theta = 90°$, $\cos\theta = 0$). When the angle θ is $90°$, the inclined plane is then a vertical face and there is no component of gravitational force holding the object against the plane.

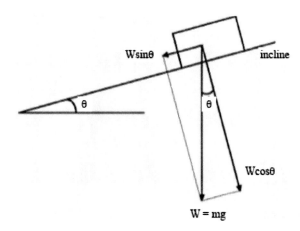

When the inclined plane is horizontal and $\theta = 0°$, the frictional force (F_{fric}) is at a maximum and defined by the coefficient of friction times the normal force. When $\theta = 0°$ the normal force is equal to the weight of the object.

The force of friction: $F_{fric} = \mu N$

Where:
F_{fric} is the force of friction in joules (J)
μ is the coefficient of friction with no units (usually between 0 and 2)
N is the normal force in joules (J)

The formula for the normal force is the weight times the cosine of the angle ($N = F\cos\theta$). Here F = weight. When $\theta = 0°$ the $\cos\theta = 1$ and therefore $N = F$. As the incline of the plane increases, the value of $\cos\theta$ decreases and therefore the level of the frictional force decreases. Less force is then required to overcome the frictional forces and the object moves with less external force required.

Levers

The principle of the lever is that a force applied in an effort to lift an object with mass is proportional to the length of the respective lever arms.

As in the illustration, a balance exists if the force (W = weight) times the distance from the fulcrum (the support or point at which a lever pivots) of individual 1 equals the force times the distance from the fulcrum of individual 2.

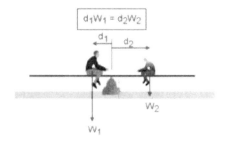

The standard equation is: $F_1d_1 = F_2d_2$

This is exactly the same as $Work_1 = Work_2$ ($Work = Fd$) and this is the expression of the conservation of energy from the input to the output.

Note that the forces could just as easily be pushing up, as long as the fulcrum is fixed to the bar in the correct orientation.

First Class Lever: Fulcrum is centered between the force and the output load

Common examples of a first class lever are a bottle opener, a crowbar, a hammer (used in pulling nails), oars, pliers, scissors, and a seesaw.

The fulcrum is located between the two forces and this is typically used to reduce the force required to lift a load.

Example 1: First Class Lever

If we have a first class lever with the following mass on the input of $m_1 = 20$ kg with the length of the input arm $d_1 = 10$ m and that of the output arm $d_2 = 1$ m calculate the mass that will be lifted at the output.

This utilizes the formula: $F_1d_1 = F_2d_2$

From $F_1d_1 = F_2d_2$, solving for F_2 gives $F_2 = \dfrac{F_1d_1}{d_2}$

Since we have the mass of the object at the input: $m_2g = \dfrac{m_1gd_1}{d_2}$

The gravity term (g) cancels and therefore $m_2 = \dfrac{m_1d_1}{d_2}$ and $m_2 = \dfrac{m_1d_1}{d_2} = \dfrac{20\text{kg}\,(10\text{m})}{1\text{m}} = 200\text{kg}$

This means that an object of 200 kg would be perfectly balanced at the output by just 20 kg at the input.

Second Class Lever: Output load is positioned between the fulcrum and the force.

Common examples of a second class lever are a door, a nutcracker, a paddle, a springboard, a wheelbarrow, and a wrench.

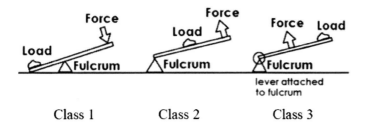

Class 1   Class 2   Class 3

The fulcrum is located at one end of the bar with the force located at the other end with the output load on the bar between the two ends.

Example 2: Second Class Lever

If we have a second class lever with the following mass on the input of $m_1 = 40$kg with the length of the input arm $d_1 = 2$m and that of the output arm $d_2 = 1$ m calculate the mass that will be lifted at the output.

This utilizes the same formula: $F_1 d_1 = F_2 d_2$

From $F_1 d_1 = F_2 d_2$, solving for F_2 gives $F_2 = \dfrac{F_1 d_1}{d_2}$

Since we have the mass of the object at the input: $m_2 g = \dfrac{m_1 g d_1}{d_2}$

The gravity term (g) cancels and therefore $m_2 = \dfrac{m_1 d_1}{d_2}$ and $m_2 = \dfrac{m_1 d_1}{d_2} = \dfrac{40\text{kg} \,(2\text{m})}{1\text{m}} = 80\text{kg}$

Third Class Lever: Force pulls at the center with the fulcrum attached at one end and the output load at the other end.

Common examples of a third class lever are a baseball bat, boat paddle, broom, fishing rod, tongs, tweezers, a shovel and a stapler.

The fulcrum is located at one end of the bar with the output load located at the other end with the input force on the bar between the two ends. Note that in this case the input force is greater than the output load but the advantage is that the output load travels a greater distance.

Example 3: Third Class Lever

If $F_1 = 20$ kg, $d_1 = 2$ m and F_2 is 8 kg then if $d_2 = 5$ m, it will balance, with more than twice the distance traveled by F_2 relative to F_1.

If we have a third class lever with the following mass on the input of $m_1 = 20$ kg with the length of the input arm $d_1 = 2$ m and the mass at the output arm $m_2 = 8$ kg calculate the length of the output arm d_2.

This utilizes the formula: $F_1 d_1 = F_2 d_2$

From $F_1 d_1 = F_2 d_2$, solving for d_2 gives $d_2 = \dfrac{F_1 d_1}{F_2}$

Since we have the mass of the object at the input: $d_2 = \dfrac{m_1 g d_1}{m_2 g}$

The gravity term (g) cancels and therefore $d_2 = \dfrac{m_1 d_1}{m_2}$ and $d_2 = \dfrac{m_1 d_1}{m_2} = \dfrac{20\text{kg}\,(2\text{m})}{8\text{kg}} = 5\text{m}$

Pulleys

Pulleys can come in a wide variety of configurations. Pulleys exist in garages, on boats, even in engines. The primary advantage of a pulley is that the load is split between the number of cables (or ropes) that hold the weight.

The formula $Work_{in} = Work_{out}$ ($F_1 d_1 = F_2 d_2$) when using pulleys is always the same. Where the mechanical advantage comes in is when the distance we pull the rope is increased which then allows us to pull with less force to raise the load. Of course the load is raised a smaller distance but we are interested in finding a way to use less force. We sacrifice by having to pull for a longer distance but we gain by being able to lift loads that are many times heavier than what we normally would be able to lift.

The Wedge

The wedge is a solid triangular shaped block. The typical use is to apply a force to the base of the triangle in order to force the tip of the triangle into a material or between two objects. The force applied at the base is amplified at the tip of the triangle because of the difference in surface area that transmits the force and the fact that the force is applied to the objects through the two triangle sides of the triangle which are usually at an extreme angle relative to the base. Therefore the amount of force applied to the base is translated to movement in between the objects against the friction of the two sides.

The Wheel and Axle

The typical example of a wheel and axle system is any bicycle, any vehicle or any belt driven manufacturing system because all of these use a wheel that rotates around an axle to guide the rotation and direction of the forces applied.

The Screw

The screw is a rotary wedge. The force is applied as a moment on the screw head and this is translated into both forward movement and circular movement into the material.

Concept Reinforcement

1. What does it mean when we say that a system is "static?"

2. What are the basics of performing a mechanical analysis?

3. What are axial and non-axial members?

4. Describe the two main types of structural support.

5. Describe the three equilibrium equations for a two dimensional system.

Section 2.4 – Centroids of Areas

Section Objective

- Describe the centroids of areas

Introduction

This section presents the concept of a center of mass of an area, called the centroid of the area. The density, which is the mass per unit area, is assumed to be uniform which allows the use of the area of the object rather than the force or load of the object in calculations. The objective is to find a centerpoint in the two-dimensional object where the amount of area, and hence load, is the same in all directions. At this point the object could be balanced on a pin. This is exactly the same as establishing a single force applied at a single location as a conceptual replacement for several forces acting at different locations within the object.

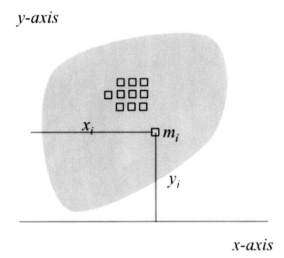

The center of mass of the area is the intersection of the center of mass about the y-axis and the center of mass about the x-axis

Center of Mass

We are going to learn how to work with objects that are not massless, meaning they have mass and will have weight in a gravitational field. This will require a different approach to the problem. First, we will examine a simple linear system and the concept of the center of mass.

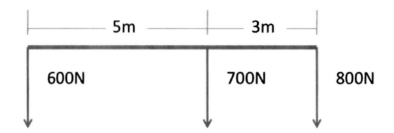

Along a massless line, we have several masses, all feeling the effects of gravity and creating forces in the down direction. Where is the center of the mass? In other words, at what point along the line would a single mass give the same effect as these various masses?

We will work with different forces, though the location of the center of mass will be in the same place.

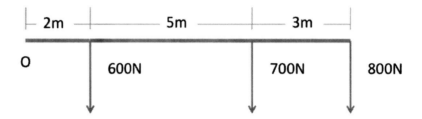

The solution is established by finding a single moment that can replace the sum of the moments at a point outside the line at point O.

$$\sum M - 600 \text{ N } (2 \text{ m}) + 700 \text{ N } (7 \text{ m}) + 800 \text{ N } (10 \text{ m})$$

To find the center location we define the x_c to be the center of mass/force.

$$x_c = \frac{\sum M}{\sum F} = \frac{600 \text{ N } (2 \text{ m}) + 700 \text{ N } (7 \text{ m}) + 800 \text{ N } (10 \text{ m})}{(600 \text{ N} + 700 \text{ N} + 800 \text{ N})} = \frac{14,100 \text{ Nm}}{2,100 \text{ N}} = 6.7 \text{ m}$$

Note that if the formula is for the center of mass. the gravity is removed from both the top and the bottom of the fraction. Therefore the solution is the same. i.e.

$$x_c = \frac{\sum mx}{\sum m} = 6.7 \text{ m}$$

Centroids of Areas

The centroid of area is the same concept as the center of mass, but applied to a two dimensional area rather than a single dimension; i.e. a line. This is also commonly called the center of gravity or center of mass of the area. This could be considered the balance point where the area would be able to balance on a point without tipping in any direction.

These calculations apply the assumption that the density of the area is uniform, which infers that the mass and, therefore the weight, is evenly distributed. These calculations include the assumption that the thickness of the area is uniform, which supports the same goal as the first assumption, which is that the mass and, therefore the weight, is evenly distributed. These assumptions allow for the possibility of standardized formulas for specific shapes.

Graphical Methods

One method to solve for the geometric center of a standard area is to draw the line bisecting each side of the area. The point of intersection is the center of mass and the geometric center.

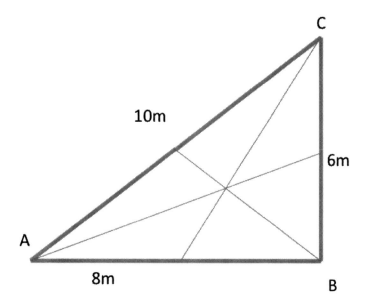

Algebraic Methods

The centroid of an area is calculated by dividing the area in one dimension into many parallel sections. Then the centroid of each area is established, keeping in mind that the center of mass should be the geometric center of the thin section. These sections are then added together and divided by the total area. This is done in a similar way to the calculation for the center of mass, which is the summation of the various centers of mass all divided by the total mass. Since the mass distribution in an area is uniform, the sum of the area is equivalent to the sum of the mass (or weight) for each section and for the total area.

Note that because the mass distribution is uniform, mass can be removed from the equations for centroids of area.

Advanced Methods

Just like with the calculation for the center of mass on a one-dimensional line, there is a way to examine the moments of an area that is also used to define the centroid of the area. The total area is divided up into many small individual squares and the moment about each axis is determined for each of the individual squares. Each square has the same mass (and the same weight). All of the individual moments are summed about one axis. This sum is divided by the total area (which is the equivalent of the weight or force because the density is uniform and the thickness is uniform). This provides the centroid about the single axis.

A second set of calculations is done using the same pattern, but this time around the other axis. Again, the individual moments are summed and the sum is divided by the total area (equivalent to the total weight) to get the centroid.

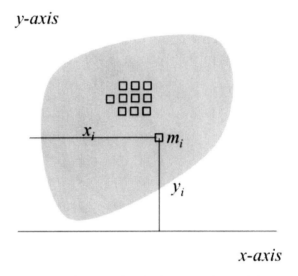

y-axis

x-axis

The center of mass of the area is the intersection of the center of mass about the *y*-axis and the center of mass about the *x*-axis

The centroid of the area is the intersection of the two centroids about the separate axes.

The calculation for the center of mass about the *y*-axis:

$$x_c = \frac{\sum_{i=1}^{n} x_i m_i}{\sum_{i=1}^{n} m_i}$$

The calculation for the center of mass about the *x*-axis:

$$y_c = \frac{\sum_{i=1}^{n} y_i m_i}{\sum_{i=1}^{n} m_i}$$

Where:
x_i = the distance to the unit of mass (m_i) from the *y*-axis in meters (m)
y_i = the distance to the unit of mass (m_i) from the *x*-axis in meters (m)
m_i = the unit of mass in kilograms (kg) as a uniform fraction of the total mass.

To calculate the centroid of an area, we instead use the area, which we noted is equivalent to the mass because of uniform density and thickness assumptions.

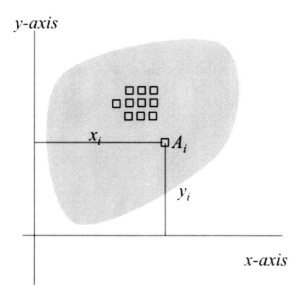

y-axis

x-axis

The centroid of the area is the intersection of the centroid about the *y*-axis and the centroid about the *x*-axis

For the centroid of an area about the *y*-axis:

$$x_c = \frac{\sum_{i=1}^{n} x_i A_i}{\sum_{i=1}^{n} A_i}$$

For the centroid of an area about the *x*-axis:

$$y_c = \frac{\sum_{i=1}^{n} y_i A_i}{\sum_{i=1}^{n} A_i}$$

Where:
x_i = the distance to the unit of area (A_i) from the *y*-axis in meters (m)
y_i = the distance to the unit of area (A_i) from the *x*-axis in meters (m)
A_i = the unit of area in square meters (m^2) as a uniform fraction of the total area.

Concept Reinforcement

1. Explain the center of mass.

2. Explain how to calculate the center of mass of a set of masses in a line.

3. Explain how to calculate the center of mass of a set of forces in a line.

4. Give the definition of the centroid of an area.

5. Describe how to determine the centroid of an area graphically.

6. Describe how to calculate the centroid of an area using algebra.

Section 2.5 – Centroids of Composite Areas

Section Objective

- Describe the centroids of composite areas

Introduction

This section presents the method to determine the centroid of a composite area. When the various pieces of the composite are geometric shapes there are tables that present formulas for locating the centroid. Once the centroid of each individual shape is located and the total area of each shape is known the calculation of the centroid for the composite of these shapes is relatively straightforward. We sum the area moments for the various shapes ($M = Ad$) and divide this sum by the total area of composite. Overall, this assumes that the mass density of all the areas is the same which allows us to use the area as a replacement for the load in calculations.

Centroid of an area

To calculate the centroid of an area, multiply the sum of the individual unit areas by the distance from the axis, and then divide by the total area. The area is equivalent to the mass because we are assuming that the area has uniform density and uniform thickness.

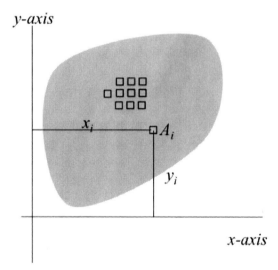

The centroid of the area is the intersection of the centroid about the *y*-axis and the centroid about the *x*-axis

For the centroid of an area about the *y*-axis:

$$x_c = \frac{\displaystyle\sum_{i=1}^{n} x_i A_i}{\displaystyle\sum_{i=1}^{n} A_i}$$

For the centroid of an area about the x-axis:

$$y_c = \frac{\sum\limits_{i=1}^{n} y_i A_i}{\sum\limits_{i=1}^{n} A_i}$$

Where:
x_i = the distance to the unit of area (A_i) from the y-axis in meters (m)
y_i = the distance to the unit of area (A_i) from the x-axis in meters (m)
A_i = the unit of area in square meters (m^2) as a uniform fraction of the total area.

The centroid of a composite area is the combination
of the centroid of the different component shapes

A Composite Area

A composite area is a set of shapes, usually geometric, that are joined together to create a single grouping. For example, using several different sized squares, the grouping will not appear like a single uniform square but can still be treated as a single total area.

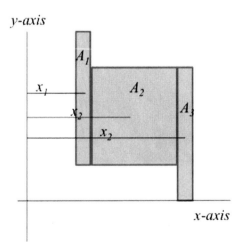

The centroid about the y-axis of a composite area is the combination
of the centroids of the different component shapes about the y-axis

Centroids of Composite Areas

The calculation for the centroid of an area about the y-axis:

$$x_c = \frac{\sum\limits_{i=l}^{n} x_i A_i}{\sum\limits_{i=l}^{n} A_i}$$

Where:

x_i = the distance to the unit of area (A_i) from the y-axis in meters (m)
y_i = the distance to the unit of area (A_i) from the x-axis in meters (m)
A_i = the area of the composite shapes in square meters (m²)

This would be $x_c = \dfrac{x_1 A_1 + x_2 A_2 + x_3 A_3}{A_1 + A_2 + A_3}$

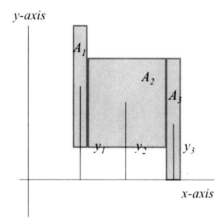

The centroid about the x-axis of a composite area is the combination
of the centroids of the different component shapes about the x-axis

The calculation for the centroid of an area about the x-axis:

$$y_c = \frac{\sum\limits_{i=l}^{n} y_i A_i}{\sum\limits_{i=l}^{n} A_i}$$

Where:

x_i = the distance to the unit of area (A_i) from the y-axis in meters (m)
y_i = the distance to the unit of area (A_i) from the x-axis in meters (m)
A_i = the area of the composite shapes in square meters (m²)

This would be $x_c = \dfrac{y_1 A_1 + y_2 A_2 + y_3 A_3}{A_1 + A_2 + A_3}$

What if one of the areas in not solid, or if one of the areas is only a space between solid areas? In this case, the empty area is treated as having no area and therefore does not participate in the calculation.

Centroids of Composite Areas with an Empty Space

For the centroid of an area about the y-axis:

$$x_c = \frac{\sum_{i=1}^{n} x_i A_i}{\sum_{i=1}^{n} A_i}$$

Where:
x_i = the distance to the unit of area (A_i) from the y-axis in meters (m)
y_i = the distance to the unit of area (A_i) from the x-axis in meters (m)
A_i = the area of the composite shapes in square meters (m²)

The resulting equation is: $x_c = \dfrac{x_1 A_1 + x_3 A_3}{A_1 + A_3}$

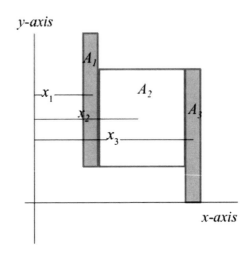

The centroid about the y-axis of a composite area is the combination of the centroids of the different component shapes about the y-axis

The calculation for the centroid of an area about the x-axis:

$$y_c = \frac{\sum_{i=1}^{n} y_i A_i}{\sum_{i=1}^{n} A_i}$$

Where:

x_i = the distance to the unit of area (A_i) from the y-axis in meters (m)
y_i = the distance to the unit of area (A_i) from the x-axis in meters (m)
A_i = the area of the composite shapes in square meters (m²)

The resulting equation is: $x_c = \dfrac{y_1 A_1 + y_3 A_3}{A_1 + A_3}$

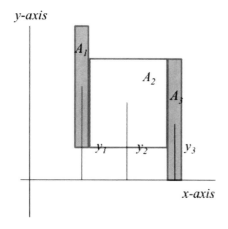

The centroid about the x-axis of a composite area is the combination of the centroids of the different component shapes about the x-axis

---

Concept Reinforcement

1. Explain the centroid of an area.

2. Explain how the centroid relates to the center of mass of an area.

3. Describe a composite area.

4. Describe the centroid of a composite area.

5. State the formula for a centroid of a composite area.

Section 2.6 – Distributed Loads

Section Objective

- Explain distributed loads

Introduction

This section presents distributed loads and the methods used to analyze systems that are experiencing distributed loads. We use the assumption that the density of the load is uniform and therefore the area can be used as a replacement for the actual load in calculations. The area of the load can then be divided into geometric shapes (as much as is possible) and the centroid of the area of each sub-area is determined. To determine the centroid of the area of the whole load we sum the area moment of the various sub-areas about a point (usually a point which is a clearly measurable distance from the point loads). This sum is divided by the total load. This defines the actual position on the beam where a point load would be applied as a replacement for the distributed load.

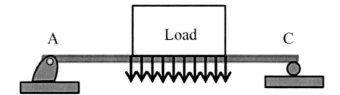

Distributed load presented as acting across the entire width of the load

Loads

A load is the term used for the weight associated with a mass above a mechanical system. When using early models for mechanical analysis, the load is a force applied at a point. It is possible for more than a single external load to be experienced by the object. These are always shown as point loads.

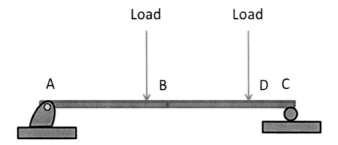

Distributed Loads

As the types of analysis become more complicated, the type of external loads experienced by an area or object become spread out and are no longer presumed to be point loads. A distributed load is more like a block with mass that rests on an object and exposes the object to a force that is spread out along the entire area of the block.

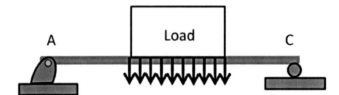

Distributed load presented as acting across the entire width of the load

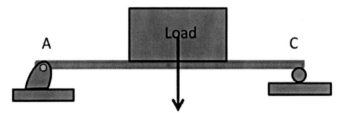

Typical Free-Body Diagram with the load presented as acting from the center of the load

The free-body diagram for a distributed load no longer shows a single force acting from the center of the object, but uses a block to represent the overall load.

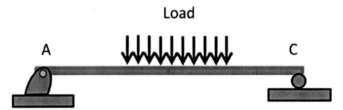

Uniformly distributed load presented as the force
acting across the entire width of the load

A distributed load shows a set of force lines acting on the object in question. The magnitude of the distributed lines of force is proportional to the fraction of the force experienced at the section. If, for instance, the load is 1,000 N and the width of the block presenting the load is 2 m, then for every 10 cm of the width the load would be 50 N. The load can also be described as 500 N per meter of width or as 5 N per centimeter.

In this particular case, a uniformly distributed load can be presented as 50 N/10 cm (or 5 N/cm or 500 N/m). The total load is calculated by multiplying this load per unit length by the length of applied load. In this case, the length of applied load is the width of the block, 2 m, which means the total load is 1,000 N. The point of application of the load is midway between the ends of the blocks. This distance is used to simplify the calculation of the moment.

We have just described how to calculate a uniform distributed load. The load can just as easily be uniformly increasing or decreasing along the width of the block.

When the load is uniformly increasing, it can be treated as a uniformly distributed load plus a triangular load. The two components have slightly different points of application when the loads are simplified to a point load.

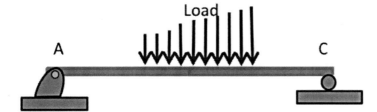

Uniformly changing distributed load presented as the force
acting across the entire width of the load

In all cases, when the mechanical system is simplified (for instance to calculate the moment), the load will be simplified to a point load acting and at a single distance.

Example 1: Distributed Load

We have a load that begins 2 m to the left from point A, which is where the system is pinned. The load is uniformly increasing across a 4 m section and begins with a 100 N/cm load on the left and ends at 1,000 N/cm load on the right. Calculate the total load and the point of application relative to point A.

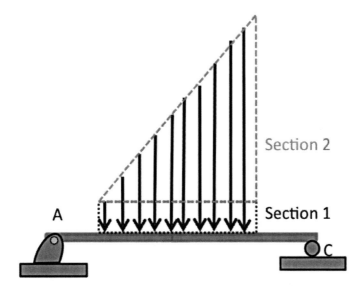

This uniformly increasing load can be divided into two parts:

1. The load/cm of 100 N/cm across the entire length of 4 m.

2. The triangular load that increases from 0 N/cm to 900 N/cm at the right.

First, determine how many cm in the 4 m section. This is 4 m(100 cm/m) = 400 cm.

The load for section 1:

This is (100 N/cm)(400 cm) = 40 kN

The load for section 2:

Here we use the formula for the area of a right triangle, which is ½ the area of the length × height. This is ½ (400 cm)(900 N/cm) =180 kN.

To determine the point of application for the load of section 1, we use the fact that the load has a rectangular shape, which places the load at ½ the width of the distributed load; i.e. 2 m in from the edge, which is a total of 4 m from point A.

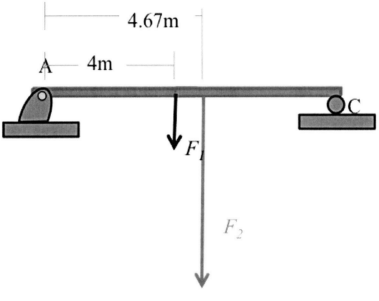

The resultant forces

To determine the point of application for the load of section 2, we use tables that have the formulas for triangular shaped loads. These tables present the formula for the centroid of an area based on a particular orientation. For a triangle, the centroid is 1/3 the distance from the base, which is the larger dimension. Therefore, 1/3 of the distance from the 900 N/cm side is 4/3 of a meter. This places the point load at a distance of 2 and 2/3 meters from the left end and therefore 4 and 2/3 m from point A.

Concept Reinforcement

1. Explain loads and the units of a load.

2. Explain distributed loads

3. Explain the difference between a uniform and a uniformly increasing distributed load.

Section 2.7 – Centroids of Volumes and Lines

Section Objective

- Describe the centroids of volumes and lines

Introduction

This section presents the centroid of a volume, which is an extension of the concept of a centroid for an area and the center of mass of a beam. As with the centroid of the area, the mass density of the volume is assumed to be uniform and therefore the volume is substituted for the mass in calculations.

The determination of the centroid of a volume is broken into three calculations, one for each plane. A planar centroid is determined by summing the volumetric moments at the various distances from the plane. This sum is divided by the total volume and the result is a distance from the plane. For instance, there is a planar centroid parallel to the plane defined by the x-y axes. ½ of the volume is above this plane and ½ below. The same exists for the other two planes described by the x-z axes and the y-z axes.

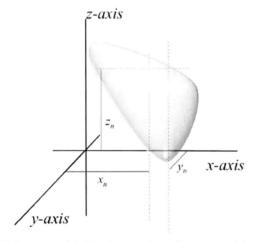

Centroid of a Volume: initially determine the centroid about each axis

Centroids

The centroid is the same as the center of mass or the center of gravity. It is calculated using dimensions of length, rather than mass or force, because several assumptions are made as part of these calculations.

- Assumption 1 is that the density, the amount of mass per unit volume is constant.

- Assumption 2 is that the thickness of the material is constant.

Using these two assumptions, the calculation for centroid in a two-dimensional system requires only the use of unit area and distance from an axis. The center of mass requires the use of a mass or force unit in the calculations.

There are technically three centroids in a two-dimensional area:

1. one about the x-axis.

2. a second about the y-axis.

3. the third is the centroid of the area that is located at the intersection of the centroids about the x and the y axes.

The centroid of an area is a point within the area at which the area will be balanced like a plate on a pin. The centroid about the x-axis is similar, but only applicable in one dimension. The area, if positioned on a line at the x-centroid would be balanced in that dimension. For the y-centroid the line would be perpendicular to the x-centroid line and the area would again be balanced.

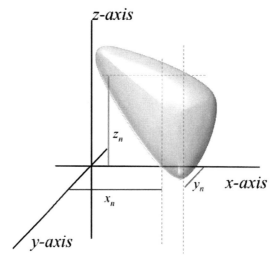

Centroid of a Volume: initially determine the centroid about each axis

Centroids of Volumes

The same formula applies in a three-dimensional system. A three-dimensional system has four centroids:

1. x-axis centroid.

2. y-axis centroid.

3. z-axis centroid.

4. the overall volumetric centroid, which is located at the intersection of the x, y and z centroids.

The centroid of a volume is a point within an object at which the 3-dimensional object will be balanced like a helium balloon in a 3-dimensional space. The centroid about the

144

x-axis is similar but only applicable in the one dimension. The volume, if positioned on a line at the *x*-centroid, will be balanced in that dimension. For the *y*-centroid, the line will be perpendicular to the *x*-centroid line and the volume will again be balanced. For the *z*-centroid, the line will be perpendicular to both the *x*-centroid line and the *y*-centroid line and the volume will again be balanced.

Calculating the Centroid of a Volume

To calculate the centroid of a volume, the sum of the individual unit volumes is multiplied by the distance from the axis. The product of this calculation is then divided by the total volume of the object. The volume is equivalent to the mass because of the uniform density assumption.

For the *x*-coordinate of the centroid of a volume:

$$x_c = \frac{\sum_{i=1}^{n} x_i V_i}{\sum_{i=1}^{n} V_i} = \frac{1}{V_T} \sum_{i=1}^{n} x_i V_i$$

This is the sum of the collection of each unit of volume multiplied by the distance of that unit volume from the *x*-axis (*y*-*z* plane). This is then divided by the total volume.

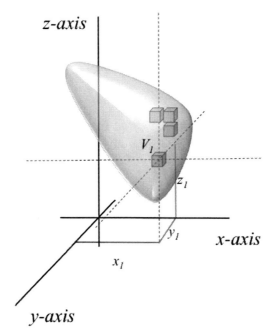

The centroid of the volume is the intersection of the centroid
about the *x*-axis, *y*-axis and the centroid about the *z*-axis

145

The calculation for the y-coordinate of the centroid of a volume:

$$y_c = \frac{\sum\limits_{i=1}^{n} y_i V_i}{\sum\limits_{i=1}^{n} V_i} = \frac{1}{V_T} \sum\limits_{i=1}^{n} y_i V_i$$

This is the sum of the collection of each unit of volume multiplied by the distance of that unit volume from the y-axis (x-z plane). This is then divided by the total volume.

The calculation for the z-coordinate of the centroid of a volume:

$$z_c = \frac{\sum\limits_{i=1}^{n} z_i V_i}{\sum\limits_{i=1}^{n} V_i} = \frac{1}{V_T} \sum\limits_{i=1}^{n} z_i V_i$$

This is the sum of the collection of each unit of volume multiplied by the distance of that unit volume from the z-axis (x-y plane). This is then divided by the total volume.

Where:
x_i = the distance to the unit of volume (V_i) from the y-z plane in meters (m)
y_i = the distance to the unit of volume (V_i) from the x-z plane in meters (m)
z_i = the distance to the unit of volume (V_i) from the x-y plane in meters (m)
V_i = the unit of volume in cubic meters (m^3) as a uniform fraction of the total volume.

Centroids of Lines

The centroid of a line is calculated in the same manner as the centroid of an area and a centroid of a volume. The line may have any shape, including linear, spiral or a random curved shape.

Therefore, the calculations for the centroid of a line appear very similar to those for the centroid of a volume. The calculations take into consideration all three dimensions within a coordinate system. The calculation of the centroid about each axis is comparable to the calculation of the moment of the line about each axis.

For the x-coordinate of the centroid of a line:

$$x_c = \frac{\sum\limits_{i=1}^{n} x_i L_i}{\sum\limits_{i=1}^{n} L_i} = \frac{1}{L_T} \sum\limits_{i=1}^{n} x L_i$$

This is the sum of the collection of each unit of length times the distance of that unit length from the x-axis (y-z plane). This is then divided by the total length.

For the *y*-coordinate of the centroid of a length:

$$y_c = \frac{\sum_{i=l}^{n} y_i L_i}{\sum_{i=l}^{n} L_i} = \frac{1}{L_T} \sum_{i=l}^{n} yL_i$$

This is the sum of the collection of each unit of length times the distance of that unit length from the *y*-axis (*x-z* plane). This is then divided by the total length.

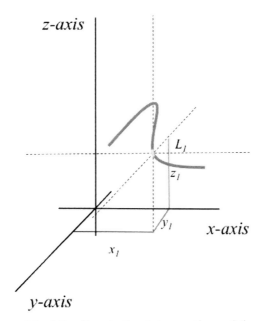

The centroid of the line is the intersection of the centroid about the *x*-axis, *y*-axis and the centroid about the *z*-axis

For the *z*-coordinate of the centroid of a length:

$$z_c = \frac{\sum_{i=l}^{n} z_i L_i}{\sum_{i=l}^{n} L_i} = \frac{1}{L_T} \sum_{i=l}^{n} zL_i$$

This is the sum of the collection of each unit of length times the distance of that unit length from the *z*-axis (*x-y* plane). This is then divided by the total length.

Where:
x_i = the distance to the unit of length (L_i) from the *y-z* plane in meters (m)
y_i = the distance to the unit of length (L_i) from the *x-z* plane in meters (m)
z_i = the distance to the unit of length (L_i) from the *x-y* plane in meters (m)
L_i = the unit of length in meters (*m*) as a uniform fraction of the total length.

Concept Reinforcement

1. Explain centroids.

2. Explain the centroid of a volume.

3. State the formula for the centroid of a volume.

4. Describe how the centroid of a volume is determined.

5. State the formula for the centroid of a line.

Section 2.8 – Pappus-Guldinus Theorems

Section Objective

- State the Pappus-Guldinus theorems

Introduction

This section presents the Pappus-Guldinus Theorems which provide algebraic formulas for determining what appear to be difficult challenges. The first theorem is focused on the external surface area of a three dimensional shell. The shell must be created by rotating a line about an axis (a line) that is at a distance from the line itself. The line can be a very complicated shape. The first Pappus-Guldinus theorem uses the concept of the three dimensional centroid of a complex line and uses this to create a formula to calculate the surface area that is defined by rotating that line 360° about the axis line. The second Pappus-Guldinus theorem takes this one step further and focuses on the volume that would be created by that same line as it is rotated about an axis. An area is described by the complex line on one side, the axis line as the other side and the top and bottom by vertical lines reaching from the axis to the complex line. The centroid of this area is calculated. Then the volume is calculated by multiplying the area described by the distance around a 360° rotation about the axis.

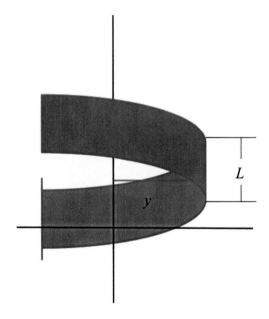

Pappus-Guldinus Theorems

The Pappus-Guldinus theorems apply the centroids of lines and areas to an interesting structure, that of a surface of revolution and a volume of revolution.

If we take a straight vertical line of 20 m and place it 1 m from an axis and then rotate that line about that axis, we draw a tube that is 20 m in height and 1 m in radius. This concept can be applied to virtually any type of line, whether it is angled or curvy relative to the central axis. These are called surfaces of revolution.

The first theorem describes a simplified formula for determining the area of a surface of revolution using the centroid of the line about that axis.

The line cannot intersect the axis of rotation.

Area of the surface of revolution: $A = 2\pi \bar{y} L$

Where:
A = the area of the surface in square meters (m^2)
$\bar{y}$ = the centroid about the x-axis in meters (m)
L = the length of the line in meters (m)

This is a restatement of the formula for area of $A = LW$.

Where:
A = the area of the surface in square meters (m^2)
W = the width of the area in meters (m)
L = the length of the area in meters (m)

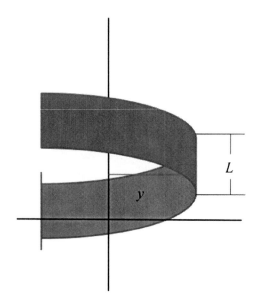

The "width" factor is defined by the centroid of the line multiplied by the arc distance. The centroid can be roughly considered to be the average distance from the axis of rotation. This equates to a radius in the formula for the circumference of a circle, which is $2\pi r$. When the line is a simple straight line, the formula is simply $A = 2\pi r L$.

When the line is much more complicated, calculations for the centroid about the axis provide a very useful shortcut.

If the area produced by the line is not a complete circle, then it is just a matter of defining the arc length as a fraction of a complete circumference.

If the line actually creates a geometric shape, such as a square, circle, triangle, or other unusual shape, then the formula from the theorem results in a surface area similar to that of a circular pipe that has either a square, circular, a triangular or an unusual profile, respectively.

The Second Pappus-Guldinus theorem

If we take a flat and straight vertical plate with an area of 20 m × 2 m, place it on an axis and then rotate that area about that axis, we draw a solid column that is 20 m in height and 4 m in diameter. This can be applied to virtually any type of area, whether it is angled, curvy or at a distance to the central axis. These are called volumes of revolution.

The second theorem describes a simplified formula for determining the volume of a volume of revolution using the centroid of the area about that axis.

The area cannot intersect the axis of rotation.

Volume of the volume of revolution: $V = 2\pi\bar{y}A$

Where:
V = the volume in cubic meters (m^3)
$\bar{y}$ = the centroid about the x-axis in meters (m)
A = the area of the surface in square meters (m^2)

This is a restatement of the formula for volume of $V = LWH$

Where:
V = the volume of the volume of revolution in cubic meters (m^3)
W = the width of the volume in meters (m)
L = the length of the volume in meters (m)
H = the height of the volume in meters (m)

If the area is produced by the $A = LH$, then the "width" factor is defined by the centroid of the area × the arc distance. The centroid could be roughly considered to be the average distance from the axis of rotation. The centroid also equates to a radius in the formula for the circumference of a circle, which is $2\pi r$. When the area is a simple flat plate, as in the explanation, the formula is simply $V = 2\pi r A$.

When the area is much more complicated, calculations for the centroid about the axis provide a very useful shortcut.

If we now take a flat and straight vertical plate with an area of 20 m × 2 m, place it 1 m from an axis and then rotate that area about that axis, we draw a thick tube that is 20 m in height, 2 m thick, 1 m in radius at the inner wall, and 3 m at the outer wall.

To solve for the volume of this thick tube, we calculate the volume of the volume defined by the outer diameter and subtract the empty volume that is described by the inner diameter.

As with the first theorem, the area can be quite complicated, can produce a fractional volume (an incomplete rotation about the axis), a thick tube that appears like a pipe that is curved around the axis.

Concept Reinforcement

1. Explain the centroid of a line.

2. Explain the centroid of an area.

3. Explain the first Pappus-Guldinus theorem.

4. Explain the second Pappus-Guldinus theorem.

5. State the formula for volume of an area of 25 m² that is revolved 1/2 of the way around an axis. The area has a centroid of 2 m.

Section 2.9 – Center of Mass, Simple Objects

Section Objectives

- Explain the center of mass of simple objects

Introduction

This section presents the concept of the center of mass for basic forms. For a one dimensional structure like a beam the center of mass is determined through the use of the moments about an axis divided by the total force. For a two-dimensional object like a flat plate there are two sets of calculations. One set of calculations divides the area into parallel strips of mass (or force) in one direction and calculates moments about an axis line in one direction. This sum, divided by the total mass (or force), results in a coordinate in one dimension for the center of mass. The second set of calculations results in the second coordinate for the center of mass in the other dimension.

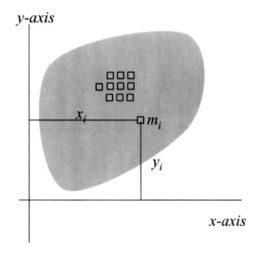

The center of mass of the area is the intersection of the center of mass about the *y*-axis and the center of mass about the *x*-axis.

For a three-dimensional object like a solid ball there are three sets of calculations. One set of calculations divides the volume into parallel 3-D plates of mass (or force). This is as if the volume was flattened to a plate and the slices of mass (or force) appear like a set of parallel forces. Using these forces the moments about an axis in one direction are calculated. This sum, divided by the total mass (or force), results in the coordinate for the center of mass in one dimension. The second and third sets of calculations result in the other two coordinates for the center of mass in the other dimensions.

Center of Application of External Forces

We will now work with objects that are not massless, meaning they have mass and, therefore, weight in a gravitational field. This will require a different approach to the problem. First, we examine a simple linear system and the concept of the center of mass.

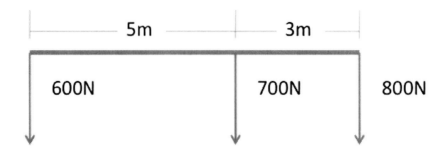

We have several masses along a massless line, all feeling the effects of gravity and creating forces in the down direction. Where is the center of the mass? This is interpreted to mean where along the line would a single mass give the same effect as these various masses?

We will work with forces in this section, even though the location of the center of mass will be in the same place (as the center of force).

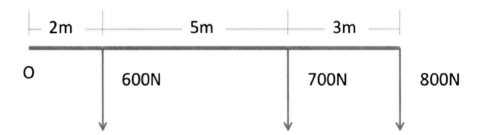

The solution is determined by finding a single moment that can replace the sum of the moments at a point outside the line at point O.

$$\sum M - 600 \text{ N } (2 \text{ m}) + 700 \text{ N } (7 \text{ m}) + 800 \text{ N } (10 \text{ m})$$

To find the center location, we define the x_c to be the center of mass/force.

$$x_c = \frac{\sum M}{\sum F} = \frac{600 \text{ N } (2 \text{ m}) + 700 \text{ N } (7 \text{ m}) + 800 \text{ N } (10 \text{ m})}{(600 \text{ N} + 700 \text{ N} + 800 \text{ N})} = \frac{14,100 \text{ Nm}}{2,100 \text{ N}} = 6.7 \text{ m}$$

Note that if the formula is for the center of mass, the gravity is removed from the top and the bottom of the fraction, therefore the solution is the same. i.e. $x_c = \frac{\sum mx}{\sum m} = 6.7 \text{ m}$.

Center of Mass

The center of mass of an object is the centroid of its mass.

Center of Mass of a Line

In one dimension along a line with two separate concentrations of mass, the center of mass is:

$$\text{Center of Mass: } x_c = \frac{\displaystyle\sum_{i=1}^{n} m_i x_i}{\displaystyle\sum_{i=1}^{n} m_i}$$

The individual point concentrations of mass are measured from a reference point outside the line. Each point concentration of mass is multiplied by its distance from the reference point. These values are summed and then divided by the total mass. The result is a single mass located at a single point along the line.

For example for two point concentrations of mass the formula is:

$$\text{Center of Mass: } x_c = \frac{m_1 x_1 + m_2 x_2}{m_1 + m_2}$$

Center of Mass of an Area

In an area in two dimensions with mass throughout the area, the center of mass is also calculated for the centroid about each axis, the x-axis and the y-axis:

$$\text{Center of Mass of the } x\text{-axis: } x_c = \frac{\displaystyle\sum_{i=1}^{n} m_i x_i}{\displaystyle\sum_{i=1}^{n} m_i}$$

$$\text{Center of Mass of the } y\text{-axis: } y_c = \frac{\displaystyle\sum_{i=1}^{n} m_i y_i}{\displaystyle\sum_{i=1}^{n} m_i}$$

The individual area concentrations of mass are measured from a reference point outside the area. This is done for all the area concentrations of mass with respect to the x-axis. Each area concentration of mass is multiplied by its distance from the reference point (usually where $x = 0$). These values are summed and then divided by the total mass. The result is a single mass located at a single point along the x-axis.

This is then repeated for along the y-axis with the end result being a single mass located at a single point along the y-axis.

The intersection of the line on the x-axis at the center of mass of the x-axis and the similar line at the center of mass of the y-axis is the center of mass of the area, the centroid of mass. This is a point at which the area can be balanced on a pin.

In a volume in three dimensions with mass throughout the volume, the center of mass is also calculated for the centroid about each axis, the x-axis, y-axis and the z-axis:

Center of Mass of the x-axis: $x_c = \dfrac{\displaystyle\sum_{i=1}^{n} m_i x_i}{\displaystyle\sum_{i=1}^{n} m_i}$

Center of Mass of the y-axis: $y_c = \dfrac{\displaystyle\sum_{i=1}^{n} m_i y_i}{\displaystyle\sum_{i=1}^{n} m_i}$

Center of Mass of the z-axis: $z_c = \dfrac{\displaystyle\sum_{i=1}^{n} m_i z_i}{\displaystyle\sum_{i=1}^{n} m_i}$

The individual volume concentrations of mass are measured from a reference plane outside the object. This is done for all the volume concentrations of mass with respect to the x-axis. Each volume concentration of mass is multiplied by its distance from the reference plane (usually where $x = 0$). These values are summed and then divided by the total mass. The result is a single mass located at a single plane parallel to the y-z plane along the x-axis.

This is then repeated for along the y-axis with the end results being a single mass located at a single plane parallel to the x-z plane along the y-axis.

This is then repeated for along the z-axis with the end results being a single mass located at a single plane parallel to the x-y plane along the z-axis.

The intersection of the plane along the x-axis at the center of mass, the plane along the y-axis at the center of mass, and the plane along the z-axis at the center of mass is the center of mass of the volume, or the centroid of mass.

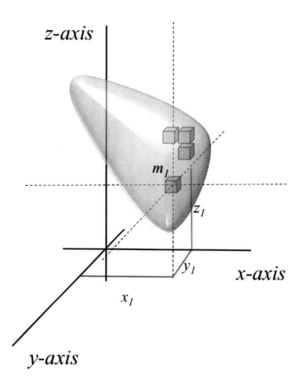

The center of mass of the volume is the intersection of the center of mass about the x-axis, the center of mass about the y-axis and the center of mass about the z-axis.

Concept Reinforcement

1. Explain the center of mass in one dimension.

2. Explain the center of mass in two dimensions.

3. Explain the center of mass in three dimensions.

4. Describe the difference between the centroid and the center of mass for an area.

5. Describe the difference between the centroid and the center of mass for a volume.

Section 2.10 – Center of Mass of Composites

Section Objective

- Explain the center of mass of composite objects

Introduction

This section presents the concept of the center of mass for a composite object. The coordinate along each axis is determined as with the center of mass determination along a one-dimensional beam. The moment of each of the parts of the composite is determined. These moments are summed and then divided by the total load of the set of parts. This results in one coordinate in one dimension. For a two dimensional object a second set of calculations are processed to obtain the coordinate in the second dimension. For a three dimensional object a third set of calculations is performed to obtain the coordinate for the third dimension.

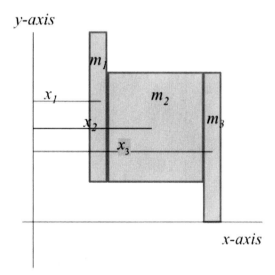

The center of mass about the *y*-axis of a composite area

Center of Mass

The center of mass of an object is the centroid of its mass.

A Composite Area

A composite area is a set of shapes, usually geometric, that are joined together to create a single grouping. For example, if you combine several different sized squares, the grouping will not appear like a single uniform square, but can still be treated as a single total area.

The calculation for the center of mass (the centroid of mass) of an area about the *y*-axis:

$$x_c = \frac{\sum\limits_{i=1}^{n} x_i m_i}{\sum\limits_{i=1}^{n} m_i}$$

Where:
x_i = the distance to the unit of mass (m_i) from the *y*-axis in meters (m)
m_i = the mass of the composite shapes in kilograms (kg)

The resulting formula is: $x_c = \dfrac{x_1 m_1 + x_2 m_2 + x_3 m_3}{m_1 + m_2 + m_3}$

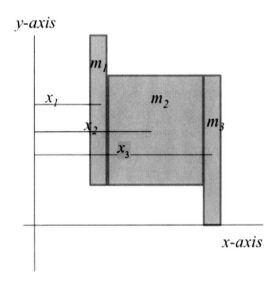

The center of mass about the *y*-axis of a composite area

The calculation for the center of mass about the *x*-axis: $y_c = \dfrac{\sum\limits_{i=1}^{n} y_i m_i}{\sum\limits_{i=1}^{n} m_i}$

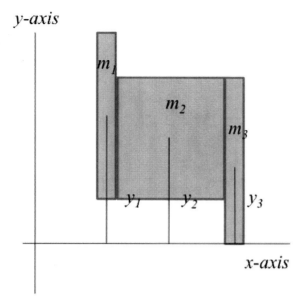

The center of mass about the x-axis of a composite area

Where:

y_i = the distance to the unit of mass (m_i) from the x-axis in meters (m)

m_i = the mass of the composite shapes in kilograms (kg)

The resulting formula is: $y_c = \dfrac{y_1 m_1 + y_2 m_2 + y_3 m_3}{m_1 + m_2 + m_3}$

What if one of the areas in not solid, or if one of the areas is only a space between solid areas? In this case the empty area is considered to have no area and therefore does not participate in the calculation.

Center of Mass of Composite Areas with an Empty Space

The formula for calculating the center of mass (centroid of mass) of an area about the y-axis:

$$x_c = \frac{\sum\limits_{i=1}^{n} x_i m_i}{\sum\limits_{i=1}^{n} m_i}$$

Where:

x_i = the distance to the unit of mass (m_i) from the y-axis in meters (m)

m_i = the mass of the composite shapes in kilograms (kg)

The resulting equation is: $x_c = \dfrac{x_1 m_1 + x_3 m_3}{m_1 + m_3}$

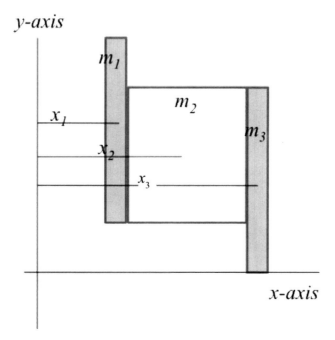

The center of mass about the *y*-axis of a composite area is the combination of the center of mass of the different component shapes about the *y*-axis

For the center of mass about the *x*-axis:

$$y_c = \frac{\sum_{i=1}^{n} y_i m_i}{\sum_{i=1}^{n} m_i}$$

Where:

y_i = the distance to the unit of mass (m_i) from the *x*-axis in meters (m)

m_i = the mass of the composite shapes in kilograms (kg)

The resulting equation is: $y_c = \dfrac{y_1 m_1 + y_3 m_3}{m_1 + m_3}$

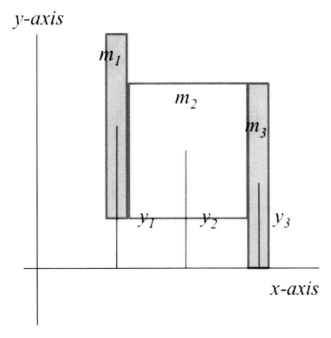

The center of mass about the *x*-axis of a composite area is the combination of the center of mass of the different component shapes about the *x*-axis.

Concept Reinforcement

1. Explain the center of mass of an area.

2. Explain how the center of mass relates to the centroid of an area.

3. Describe a composite area.

4. Describe the center of mass of a composite area.

5. State the formula for the center of mass of a composite area.

Section 2.11 – Moment of Inertia

Section Objective

- Describe the concept of moment of inertia

Introduction

This section presents the idea of the moment of inertia. With linear motion the mass of the object causes resistance to movement. For rotational motion the moment of inertia causes resistance to rotary movement. The resistance is greater if the mass of the object is further from the axis of rotation.

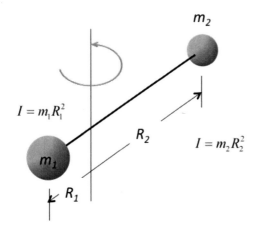

Mass is concentrated in two spheres

A Definition of Inertia

When an object is moving at a constant velocity, there is a tendency for it to continue in that direction. Can you imagine a scenario where you and your book are traveling on a train and the book suddenly does not move at the same speed with you and the train? It has inertia because it is moving at a constant velocity provided by the train. But if the train has to suddenly stop, then the inertia to move at that constant velocity which you and the book have will cause both you and the book to lurch forward, relative to the train. In the same sense, when the train makes a strong turn the inertia to move at that constant velocity (straight forward) which you and the book have will cause both you and the book to lurch to the side of the train opposite the direction of the turn.

Inertia also exists when an object is stationary. There is a tendency for it to continue to not move. This explains why a lot of force is required initially to get a car moving. This also explains why it is impossible to get your brother or sister off the couch when they settle in.

Inertia is a tendency. It is a measure of how difficult it is to change what an object is doing.

1. *Physics.* The tendency of a body to resist acceleration; the tendency of a body at rest to remain at rest or of a body in straight line motion to stay in motion in a straight line unless acted on by an outside force.

2. Resistance or disinclination to motion, action, or change: *the inertia of an entrenched bureaucracy.*

The use of the term "inertia" to refer to the difficulty of changing the direction of a bureaucracy or sometimes even an individual is an apt metaphor for the physics definition.

Applying the Concept of Inertia

When a body is not moving it is inert (has inertia). It will not move without some force causing it to move. Force is defined as the acceleration of a mass.

If the body is initially at rest and it starts to move, there is a change in velocity, which is by definition, acceleration. Hence, a force has acted on the body to get it to move. And by definition, an acceleration of the body was caused by that force. Inertia resists this movement.

A ball lying on the floor will not move unless someone or something causes it to move. That movement, or change in velocity, is acceleration. The ball has mass and hence a force ($F = ma$) has acted on the ball.

If a body is moving at a constant velocity, there is no change in velocity from one time to the next. Hence, there is no acceleration, and therefore no force acting on the body. This moving body also has inertia. In order for it to accelerate, which by definition, is to change its velocity, a force must be applied to the body.

A ball is rolling at a constant velocity. If it is on flat ground, there are forces acting on it to slow it down, hence there is acceleration in the opposite direction to the velocity and therefore a force is acting on the ball.

While riding in a car at a constant velocity, you feel no pressure other than gravity holding you to the seat. No horizontal forces are felt. Suddenly the driver decides to change the velocity. A change in velocity is acceleration. When this happens, a force is applied to the mass of the car to make it accelerate to a faster (or slower) velocity.

What happened to the passengers? They were moving along nicely at a constant velocity and suddenly a force was applied to the car. The passengers did not have that same acceleration, that same exact force, acting on them. Therefore, they felt movement of the car relative to their position. They were no longer in what is called an "inertial frame of reference."

An inertial frame of reference is a frame of reference in which the motion of a particle not subject to forces is a straight line.

The velocity of the car was changing, so the velocity of the passengers was then being forced to change by the force of the car around them acting on them. As a result, the passengers felt force from the car.

While the passengers felt like they were "thrown in the seat" when the car "lurched" forward, what actually happened was that the car jumped towards the passengers. As the car jumped toward the passengers, it pushed on them so they would start to move at the higher velocity the car had achieved. When braking the car, the opposite is true. The passengers have inertia to continue in their forward movement but the car has been forced to stop. The braking force acts against the inertia of the car.

Moment of Inertia

The moment of inertia is a measure of an object's resistance to changes in its rotational rate. The moment of inertia is rotational inertia. This is often equated with mass in a linear movement framework, beginning with the resistance to change in its rate of linear movement. As a result, the moment of inertia is also called the angular mass.

Linear Dynamics		Angular Dynamics	
Term	**Formula**	**Term**	**Formula**
Mass	m	Inertia	I
Velocity	v	Angular velocity	ω
Acceleration	a	Angular acceleration	α
Newton's 2nd Law: Force	$F=ma$	Newton's 2nd Law: Torque or Moment	$\tau = I\alpha$
Momentum	$p = mv$	Angular momentum	$L = I\omega$
Kinetic Energy	$K = \frac{1}{2}mv^2$	Angular Kinetic Energy	$K = \frac{1}{2}I\omega^2$
Work-Energy	$Fd = \Delta\left(\frac{1}{2}mv^2\right)$	Angular Work-Energy	$\tau\theta = \Delta\left(\frac{1}{2}I\omega^2\right)$

Concept Reinforcement

1. Explain inertia.

2. Explain linear inertia.

3. Explain the difference between linear and rotational inertia.

4. Explain the difference between the formula for linear kinetic energy with that of angular kinetic energy.

Section 2.12 – Moments of Simple Objects

Section Objective

- Explain moments of inertia of simple objects

Introduction

This section presents the moments of inertia of simple objects that are geometric shapes. The general formula for the moment of inertia is the product of the mass of the object and the radial distance between the axis of rotation and the center of mass. For any geometric shape the formula for the moment of inertia is established while for complicated shapes the formula for the center of mass as well as the moment of inertia require more advanced formulas.

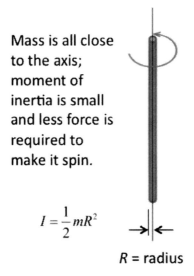

Mass is all close to the axis; moment of inertia is small and less force is required to make it spin.

$$I = \frac{1}{2} mR^2$$

R = radius

Moment of Inertia

The moment of inertia is calculated as the mass × the square of the distance to the center of mass.

Moment of Inertia: $I = mr^2$

Where:
I = the moment of inertia in kilogram-meters squared (kg-m^2)
m = mass in kilograms (kg)
r = radial distance from the axis of rotation to the center of mass in meters (m)

This is the expression for a single point mass but can be expanding to include many point masses:

Moment of Inertia of many point masses: $I = \sum_{i=1}^{n} m_i r_i^2$

Where:
I = the moment of inertia in kilogram-meters squared (kg-m²)
m = mass in kilograms (kg)
r = radial distance from the axis of rotation to the center of mass in meters (m)

In general, the formula for the moment of inertia will be some fraction of the above equation primarily because most objects are not a point mass at a distance r from the axis of rotation. Therefore, the formula $I = mr^2$ gives the magnitude that is the most that the moment of inertia will be.

Moment of Inertia of Simple Objects

For many geometric shapes the formulas have been defined. For more complicated shapes additional calculations may be needed if the complicated shape cannot be decomposed into several different common geometry shapes.

Object	Figure	Formula
Solid cylinder		$I = \dfrac{1}{2}mr^2$ Where: m = total mass in kilograms (mg) r = outer radius in meters (m)
Hoop (not solid plate)		$I = mr^2$
Solid sphere		$I = \dfrac{2}{5}mr^2$
Rod		$I = \dfrac{1}{12}ml^2$

Concept Reinforcement

1. Explain the moment of inertia.

2. Explain the difference between mass and the moment of inertia.

3. State the formula for the moment of inertia of a solid sphere.

4. Calculate the moment of inertia for a solid sphere with a mass of 300 kg and an outer radius of 0.320 m.

Section 2.13 – Rotated and Principal Axes

Section Objective

- Describe rotated and principal axes

Introduction

This section presents the concept of realigning the coordinate system to the object. The purpose of realigning a set of coordinates is to find an angle at which the moments of inertia of the object are at a minimum. This new set of axes at this angular position is called the principal axes and the minimum moments of inertia are called the principal moments of inertia.

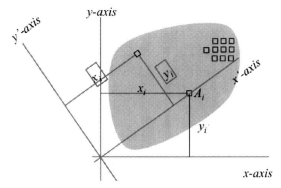

A rotated coordinate system shown over the standard coordinate system

Cartesian Coordinate System

All objects exist in the Cartesian coordinate system, which includes x, y, and z axes. This can be presented in a variety of ways but there are two generally common presentations. One has the x-axis directed towards the lower right and the y-axis is towards the upper right and the z-axis is straight up.

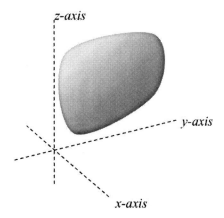

One common pattern for the Cartesian coordinate system

The other common system places the *x*-axis horizontal, the *y*-axis advancing down to the left and the *z*-axis straight up.

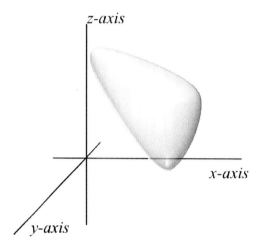

One common pattern for the Cartesian coordinate system

Though the object may be in the same relative location, the coordinate system is changed in our view. In both cases, the moments of inertia around the three axes will be the same; i.e. the moment of inertia about the *x*-axis will have the same magnitude in both presentations of the coordinate system.

Equivalent coordinate systems: $I_{x1} = I_{x2}$

There are times when the entire coordinate system is shifted by a consistent defined amount or rotated to a new orientation with respect to the object. Once the coordinate system is shifted, the moment of inertia about each of the axes in this new position will be different from that determined using the standard coordinate system.

Shifted coordinate systems: $I_{x1} \neq I_{x2}$

For some engineering applications, it is necessary to determine the moments of inertia of areas or volumes from different points of view, in the context of different coordinate systems. At other times, it is desired to determine the position of a coordinate system that results in the moments of inertia being either minimized or maximized. These adjusted coordinate systems relate the Principal axes and the Rotated axes.

Rotated Axes

As mentioned above the axes can be rotated in any orientation. The general terminology is that these are "rotated" axes without a specific endpoint for the amount of rotation.

As an example, we examine the area as shown in a two-dimensional coordinate system. The red coordinate system has been rotated by some angle (θ) to a new position.

The moment of inertia about the *x*-axis is not the same as the moment of inertia about the *x*'-axis. The moment of inertia about the *y*-axis is not the same as the moment of inertia about the *y*'-axis.

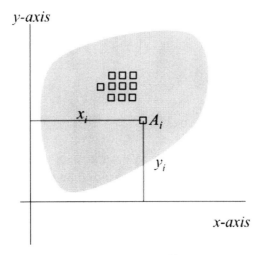

An object in an *x-y* Coordinate system

We can estimate that the moment of inertia of this object about the *y'*-axis is less than that about the *y*-axis because the mass of the object is the same but the distances from the *y'*-axis are less than those from the *y*-axis.

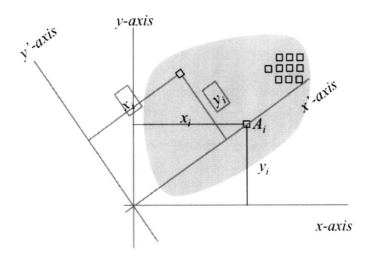

A rotated Coordinate system show over the standard coordinate system

Eventually the rotation will lead to a minimum or a maximum moment of inertia about that axis.

At that point, the term used for the coordinate system in that position is Principal Axes.

Principal Axes

The Principal Axes occur when the coordinate system is rotated to a point where the moment of inertia is minimum about one of the axes.

At this point, the moment of inertia is called the principal moment of inertia.

Principal moments of inertia: $= \dfrac{I_x + I_y}{2} \pm \sqrt{\left(\dfrac{I_x - I_y}{2}\right)^2 + \left(I_{xy}\right)^2}$

Practical Principal Axes

How does this relate to practical issues?

Usually, it is assumed that the coordinate system is a set of Principal Axes, but this is not always the case. For example, a wheel rotates around an axle. When the wheel is mounted properly, a minimum amount of energy is required to maintain rotation of the wheel. The axle is an example of a Principal Axis.

If the wheel is not mounted properly and it wobbles, the amount of energy required to maintain the rotational movement is not at a minimum, which means that the axle is no longer a Principal axis.

Concept Reinforcement

1. Explain a coordinate system.

2. Explain how the moment of inertia is affected by the location of the coordinate system.

3. Explain the difference between a translated coordinate system and a rotated coordinate system.

4. Describe how a rotational coordinate system is used with moments of inertia.

5. Explain principal axes.

Section 2.14 – Parallel Axis Theorem

Section Objective

- Explain the parallel axis theorem.

Introduction

This section presents the parallel axis theorem. Any moment of inertia for an object about an axis parallel to the axis of rotation at the center of mass of the object will be larger than the moment of inertia of the object at that center of mass.

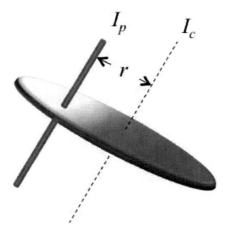

Parallel Axis Theorem

Minimum Moments of Inertia

Within the dimensions of an object, there is a position where the moment of inertia about each axis is at a minimum. This is the center of mass of the area or object about that specific axis. This will rarely be the minimum moment of inertia about all the axes at the same time, because the minimum moment of inertia about all axes at the same time would only occur with a circle area in two dimensions or a spherical volume in 3-dimensions.

Therefore, there will normally be a minimum moment of inertia in only one dimension about a specific axis. Because the mass, area or volume of the object does not change, this minimum moment of inertia will occur where the distances are at a minimum, somewhere near the center of the object.

If we examine the moment of inertia about an axis that is in the same direction, but in a slightly different location, we can guarantee that the moment of inertia will be larger. This is the basis of the parallel axis theorem.

The parallel axis theorem gives a formula for determining the moment of inertia about any parallel axis if the mass and the minimum moment of inertia are known.

The minimum moment of inertia is the moment about an axis through the center of mass of the object.

Parallel Axis Theorem: $I_p = I_c + mr^2$

Where:
I_p = the moment of inertia about a parallel axis in kilogram-meter squared (kg-m^2)
I_c = the minimum moment of inertia about an axis in kilogram-meter squared (kg-m^2)
m = mass in kilograms (kg)
r = the distance from the axis with a minimum moment of inertia to the new parallel axis in meters (m).

The component added (mr^2) is the formula for the moment of inertia of a point mass at a distance r from the axis.

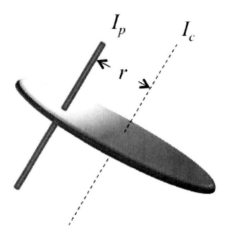

Parallel Axis Theorem

Therefore, the moment of inertia about a parallel axis is the moment of inertia at the center of mass plus the entire mass treated as a point mass at the center of mass of the object.

Example: Parallel Axis Theorem

An object with mass of 300 kg has a moment of inertia at its center of mass of 1,000 kgm^2.

Determine the moment of inertia at a parallel axis at a distance 20 cm from the center of mass.

Using the formula $I_p = I_c + mr^2$, we have I_p = 1,000 kgm^2 + (300 kg) (0.20 m^2)

This can be simplified to: I_p = 1,000 kgm^2 + (12 kgm^2) = 1,012 kgm^2

Concept Reinforcement

1. Explain the center of mass of an object.

2. Explain the minimum moment of inertia.

3. Explain the formula for the minimum moment of inertia.

4. State the formula for the parallel axis theorem.

5. An object with mass 300 kg has a moment of inertia at its center of mass of 1,000 kgm^2. Determine the moment of inertia at a parallel axis at a distance 2.0 m from the center of mass.

Section 2.15 – Analysis Application

Section Objective

- List applications of mechanical engineering

Introduction

This section presents various applications of the methods used to analyze forces and moments; i.e. the method of joints and the method of sections. The concept of a broader category of structural elements called frames and machines extends the category. Centroids of areas, centroids of volumes are examined. The center of mass and the moment of inertia are explained along with methods to analyze the moment of inertia.

Frame Analysis

In the analysis of the forces in a frame that is in mechanical equilibrium there are three typical steps:

Create a free-body diagram

Identify all the forces and identify the x and y components of each force

Solve the three equilibrium equations:

A. $\sum F_x = 0$

B. $\sum F_y = 0$

C. $\sum M_z = 0$

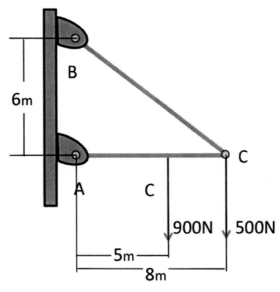

A frame with pin connections in all three joints

Frame Problem

Two members are attached to a wall in a triangular pattern. One member is 8 m long and mounted horizontally. The other member is longer, mounted at the wall at 6 m higher, and is connected by a pin to the other end of the horizontal member. At the point C, where the two members join, is a load of 500 N. At a point B, 5 m from the wall pin, is a second load of 900 N. Determine the loads at the pinned joints on the wall and produce a full free body diagram.

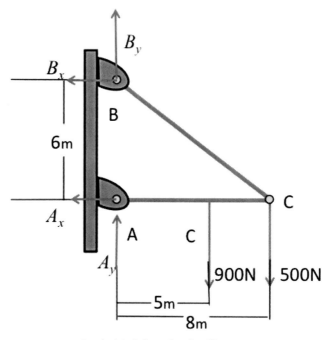

An initial free-body diagram

Step 1: Create a free-body diagram of all external forces and all dimensions, noting angles where available.

First, we show the components of the forces in all locations, specifically the two joints at the wall. Note that we technically only have three equilibrium equations and, therefore, can only solve for 3 unknown forces. At first, it appears we have four unknown forces A_x, A_y, B_x and B_y. However, because we can derive all the dimensions, we can derive the angles. Therefore, the two forces at point B plus the angle provide additional equations to help with the solution.

The equations of equilibrium are:

$$\sum F_x = = -A_x - B_x$$

$$\sum F_y = 0 = A_y + B_y - 900 \text{ N} - 500 \text{ N}$$

$$\sum M_a = 0 = 5 \text{ m } (900 \text{ N}) + 8 \text{ m } (500 \text{ N}) - 6 \text{ m } (B_x)$$

$B_x = B\cos37$ and $B_y = B\sin37$

Therefore:

$$\sum F_x = 0 = -A_x - B_x = -A_x - B\cos37$$

$$\sum F_y = 0 = A_y + B_y - 900\text{ N} - 500\text{ N} = A_y + B\sin37 - 900\text{ N} - 500\text{ N}$$

$$\sum F_y = 0 = 5\text{ m }(900\text{ N}) + 8\text{ m }(500\text{ N}) - 6\text{ m }(B_x) = (4{,}500\text{ Nm}) + (4{,}000\text{ Nm}) - 6\text{ m}(B\cos37)$$

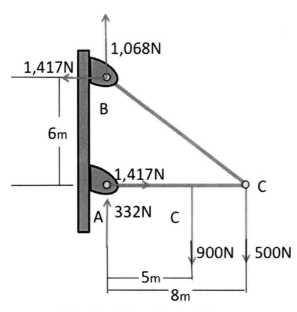

The final Free-Body Diagram

We can immediately solve for B_x from the equilibrium equation for the moments.

$$B\cos37 = \frac{(4{,}500\text{Nm}) + (4{,}000\text{Nm})}{6\text{m}} = 1417\text{N}$$

And

$$B = \frac{(4{,}500\text{Nm}) + (4{,}000\text{Nm})}{(6\text{m}\cos37)} = 1{,}774\text{N}$$

From here, we can also solve for A_x, which is the opposite value of B_x from the equation for the x-direction: $-A_x = B\cos37 = -1{,}417$ N

This is negative, therefore the direction shown in the figure is incorrect: the force is in the opposite direction and the member AC is in compression with the force to the right.

Solving for B_y: $B_y = B\sin37 = 1{,}774$ N $(\sin37) = 1{,}068$ N

And finally solving for A_y: $A_y = 900$ N $+ 500$ N $- B\sin37 = 1{,}400$ N $- 1{,}068$ N $= 332$ N

Centroids of Areas

To calculate the centroid, we use individual units of the area along with the length from the axis to determine the centroid.

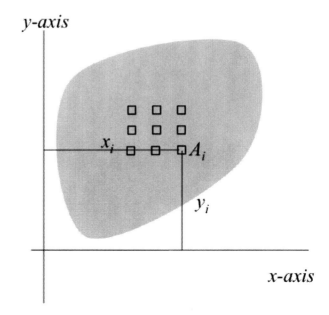

The centroid of the area is the intersection of the centroid about the y-axis and the centroid about the x-axis.

The equation used to calculate the centroid of an area about the *y*-axis:

$$x_c = \frac{\sum\limits_{i=1}^{n} x_i A_i}{\sum\limits_{i=1}^{n} A_i}$$

The equation used to calculate the centroid of an area about the *x*-axis:

$$y_c = \frac{\sum\limits_{i=1}^{n} y_i A_i}{\sum\limits_{i=1}^{n} A_i}$$

Where:
x_i = the distance to the unit of area (A_i) from the *y*-axis in meters (m)
y_i = the distance to the unit of area (A_i) from the *x*-axis in meters (m)
A_i = the unit of area in square meters (m²) as a uniform fraction of the total area.

Example 1

To approximate this work, we can assume the area has been divided up into 9 equal areas, which are uniformly distributed in a 3 × 3 pattern throughout the object. Each area is 2 m². Each area is 80 cm apart in the x-direction and the y-direction. The distance from the axis point of $x = 0$ to the first area is 50 cm. The distance from the axis point of $y = 0$ to the first area is 120 cm.

Determine the centroid about the x-axis, the y-axis and the centroid of the area.

The centroid about the y-axis is defined at a position on the x-axis. We can examine one row of areas.

$$x_c = \frac{0.50\text{m}\left(2\text{m}^2\right) + 1.30\text{m}\left(2\text{m}^2\right) + 2.10\text{m}\left(2\text{m}^2\right)}{3\left(2\text{m}^2\right)}$$

$$x_c = \left(\frac{1\text{ m}^3 + 2.6\text{ m}^3 + 4.2\text{ m}^3}{6\text{ m}^2}\right) = \frac{7.8\text{ m}^3}{6\text{ m}^2} = 1.3\text{ m}$$

In relation to the y-axis, the other two rows will provide the exact same results because the area is uniformly distributed. Therefore, even if we were to sum all nine units of area to determine the centroid about the y-axis, the result would be the same, that the $x_c = 1.3$ m

The center of mass about the x-axis is defined at a position on the y-axis. We can examine one row of point masses.

$$y_c = \frac{1.20\text{m}\left(2\text{m}^2\right) + 2.0\text{m}\left(2\text{m}^2\right) + 2.80\text{m}\left(2\text{m}^2\right)}{3\left(2\text{m}^2\right)}$$

$$y_c = \left(\frac{2.4\text{ m}^3 + 4.0\text{ m}^3 + 5.6\text{ m}^3}{6\text{ m}^2}\right) = \frac{12\text{ m}^3}{6\text{ m}^2} = 2.0\text{ m}$$

In relation to the x-axis, the other two rows will provide the exact same results because the area is uniformly distributed. Therefore, even if we were to sum all nine units of area to determine the centroid about the x-axis, the result would be the same, that the $y_c = 2.0$ m

The centroid of the area is the intersection of the two centroids. This is at the (x, y) coordinate of (1.3, 2.0).

In a two-dimensional area with mass throughout the area, the center of mass is also calculated about each axis: the x-axis and the y-axis:

Center of Mass of the x-axis: $x_c = \dfrac{\sum\limits_{i=1}^{n} m_i x_i}{\sum\limits_{i=1}^{n} m_i}$

Center of Mass of the y-axis: $y_c = \dfrac{\sum\limits_{i=1}^{n} m_i y_i}{\sum\limits_{i=1}^{n} m_i}$

The individual area concentrations of mass are measured from a reference point outside the area. This is done for all the area concentrations of mass with respect to the x-axis.

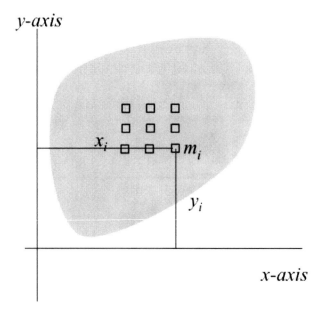

The center of mass of the area is the intersection of the center of mass about the y-axis and the center of mass about the x-axis.

Example 2

To approximate this work, we can assume we have 9 concentrations of mass uniformly distributed throughout the object in a 3×3 pattern. Each mass concentration is 50 kg. Each mass concentration is 20 cm apart in the x-direction and the y-direction. The closest mass to the point of $x = 0$ is at a distance of 40 cm. The closest mass to the point of $y = 0$ is at a distance of 30 cm.

Determine the center of mass about the x-axis, the y-axis and the center of mass of the area.

The center of mass about the y-axis is defined at a position on the x-axis. We can examine one row of point masses.

$$x_c = \frac{50\text{kg}(0.40\text{m}) + 50\text{kg}(0.60\text{m}) + 50\text{kg}(0.80\text{m})}{3(50\text{kg})}$$

$$x_c = \left(\frac{20 \text{ kgm} + 30 \text{ kgm} + 40 \text{ kgm}}{150 \text{ kg}}\right) = \frac{90 \text{ kgm}}{150 \text{ kg}} = 0.60 \text{ m}$$

In relation to the y-axis, the other two rows will provide the exact same results because the mass is uniformly distributed. Therefore, even if we were to sum all nine point masses to determine the center of mass about the y-axis, the result would be the same: $x_c = 0.6$ m

The center of mass about the x-axis is defined at a position on the y-axis. We can examine one row of point masses.

$$y_c = \frac{50\text{kg}(0.30\text{m}) + 50\text{kg}(0.50\text{m}) + 50\text{kg}(0.70\text{m})}{3(50\text{kg})}$$

$$y_c = \left(\frac{15 \text{ kgm} + 25 \text{ kgm} + 35 \text{ kgm}}{150 \text{ kg}}\right) = \frac{75 \text{ kgm}}{150 \text{ kg}} = 0.50 \text{ m}$$

In relation to the x-axis, the other two rows will provide the exact same results because the mass is uniformly distributed. Therefore, even if we were to sum all nine point masses to determine the center of mass about the x-axis, the result would be the same, that the $y_c = 0.5$ m

The center of mass of the area is the intersection of the two centers of mass. This is at the (x, y) coordinate of $(0.6, 0.5)$.

Moment of Inertia

A surface area in the shape of a hoop has a mass of 50 kg and a radius of 30 cm. Determine the moment of inertia of the hoop about the axis that is perpendicular to the plane of the circle of the hoop.

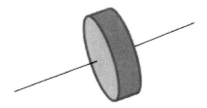

This is an application of the formula of the moment of inertia for a hoop: $I = mr^2$

Moment of inertia: $I = mr^2 = 50 \text{ kg } (0.30 \text{ m})^2 = 50 \text{ kg } (0.09 \text{ m}^2) = 4.5 \text{ kgm}^2$

Moment of Inertia

A solid sphere has a mass of 50 kg and a radius of 30 cm. Determine the moment of inertia of the circle about the central axis.

This is an application of the formula of the moment of inertia for a solid sphere: $I = \frac{2}{5}mr^2$

Moment of inertia: $I = \frac{2}{5}mr^2 = \frac{2}{5}\left(50\text{kg}(0.30\text{m})^2\right) = \frac{2}{5}50\text{kg}\left(0.09\text{m}^2\right) = 1.8\text{kgm}^2$

Parallel Axis Theorem

An object with mass 500 kg has a moment of inertia at its center of mass of 400 kg/m². Determine the moment of inertia at a parallel axis at a distance 3.0 m from the center of mass.

Using the formula $I_p = I_c + mr^2$, we have $I_p = 400$ kgm² + (500 kg) (3.0 m²)

This can be simplified to: $I_p = 400$ kgm² + (4,500 kgm²) = 4,900 kgm²

Concept Reinforcement

1. Explain frames.

2. Explain the three equations of equilibrium in a two dimensional problem.

3. Explain the four equations of equilibrium in a three dimensional problem.

4. Describe the difference between the centroid and the center of mass of an area.

5. A solid sphere has a mass of 400 kg and a radius of 750 cm. Determine the moment of inertia of the circle about the central axis.

Unit Three

Section 3.1 – Dry Friction   190

Section 3.2 – Structural Support Beams   199

Section 3.3 – Force and Moment   203

Section 3.4 – Shear Force and Bending Moment   207

Section 3.5 – Distributed Load Moment   215

Section 3.6 – Cables   225

Section 3.7 – Loads Along Straight Lines   229

Section 3.8 – Loads along Cables   235

Section 3.9 – Discrete Loads on a Cable   241

Section 3.10 – Characteristics of Liquids and Gases   249

Section 3.11 – Pressure, Center of Pressure   259

Section 3.12 – Pressure in a Stationary Liquid   263

Section 3.13 – Virtual Work   267

Section 3.14 – Potential Energy   273

Section 3.15 – Applications   281

Section 3.1 – Dry Friction

Section Objective

- Explain dry friction

Introduction

This section presents the concept of friction, a resistance force that affects virtually all surfaces. Friction comes in two types, static and kinetic friction. Static friction is the resistive force between two surfaces in contact that are not moving relative to each other. Kinetic friction is the resistive force between two surfaces in contact that are moving relative to each other. The magnitude of the force of friction is defined as a fraction of the normal force. That fraction is defined by the coefficient of friction. The direction of the force of friction is always in opposition to the direction of movement.

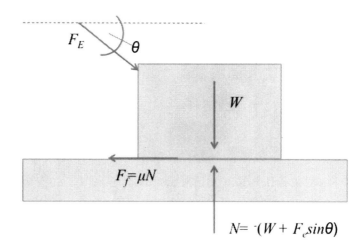

A Definition of Friction Forces

Friction is the term used to describe the force of resistance to movement. Friction exists when one surface acts against another. For example, a paperclip will continue to hold sheets of paper together because of the friction between the clip and the paper. Likewise, if those sheets of papers are between the pages in a closed book, the friction between the sheets of paper and the pages of the book holds the sheets in the book.

Climbing would be practically impossible without friction.

Technically, friction is a form of chemical adhesion and bonding between the surfaces. The molecules of the two surfaces chemically bond when the surfaces are stationary relative to each other.

These bonds break when an external force is strong enough to cause the surfaces to move relative to each other. If the surfaces again stop moving relative to each other, the molecules of the surfaces re-bond in this new position.

The roughness of a surface determines the amount of friction that will occur. This is because adhesion depends upon the amount of surface area where the two surfaces contact each other.

Friction is separated into two broad types:

- Static friction (μ_s): This when the two objects are not moving relative to each other.

- Kinetic friction (μ_k): This is when the two objects are moving relative to each other.

In all cases, static friction is greater than kinetic friction. Once the objects begin to move, friction is greatly decreased and less force is required to continue moving the objects at the same speed.

Friction in Everyday Motion

When a box on a table seems to resist your efforts at pushing it, there is friction between the bottom of the box and the surface of the table. The fact that your chair does not shift around easily and needs to be pushed to make it move shows there is friction between the feet of the chair and the floor.

In the illustration, the blue block is pulled down against the green plate by the force of gravity (W) acting on the block vertically downward. Because there is an angle of application of the external force (F_E) only a portion of the external force is acting on the block to push it towards the right. The total apparent weight of the block is therefore the weight (W) of the block plus the vertical portion of the external force ($F_E\sin\theta$) acting on the block.

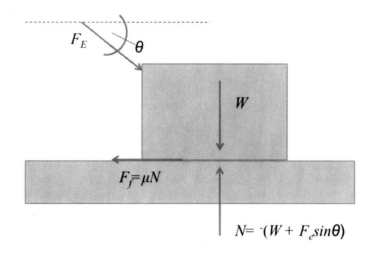

$$N = \bar{}(W + F_e\sin\theta)$$

The remaining portion of the external force ($F_E\cos\theta$) acts to push the block to the right in opposition to the forces of friction, which are acting at the interface between the block and the surface. The "Normal Force" (N) is an equal and opposite force of the surface acting against the apparent weight of the block. The normal force is equal and opposite to the sum of the force due to gravity and the external force acting at this angle θ ($N = (W + F_E\sin\theta)$).

The force of friction (F_f) is equal to the coefficient of friction (μ) between the two surfaces times the normal force. Because the normal force is a negative value, the friction force is in the opposite direction of the movement. ($F_f = \mu N$).

Friction is what causes your tires to stay connected to the road while driving. When a car is parked, it does not slide away because of the friction between the tires and the road surface. When the road and the tires are dry, the tires grip the road even at high speeds. When the road is wet or worse, oily, the frictional forces are weaker and it is more likely that the tires will slip. This is the prime reason there are so many more auto accidents in wet road conditions.

When everything is dry, the friction between solid objects is called dry friction or sliding friction. The friction between a solid and a liquid is called fluid friction.

Friction forces exist in our world at all times. There is always some amount of friction resisting movement.

In the illustration, the pink block is pulled down the inclined plane by the force of gravity (W) acting on the block vertically downward. Because there is an angle of incline, only a portion of the gravitational force is acting on the block to pull it down the incline. The actual gravitational force acting on the block down the ramp is $W\sin\theta$. The remaining portion of the gravitational force ($W\cos\theta$) acts to pull the block against the surface. The "Normal Force" (N) is an equal and opposite force of the inclined plane against the block, meaning it is equal and opposite to the force of gravity at this angle of the block against the surface ($N = W\cos\theta$).

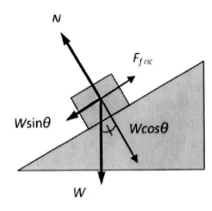

If the inclined plane were horizontal, this normal force would be directly vertical. The normal force is always perpendicular to the surface. The frictional force (F_{fric}) acts against the direction of motion at the surface between the block and the inclined plane and is proportional to the normal force.

Books "stick" to each other, the dresser drawers stick to the sides, and your plates "stick" to the table.

It would not be possible to write on paper with a pen if there were no frictional forces. Have you noticed how some types of paper are easier to write on or draw on than others? This is because the frictional forces are different between the pen or pencil and the different types of paper.

As is noted above, friction is more related to chemical adhesion between the two surfaces than to surface roughness. The amount of friction between two smooth surfaces will typically be stronger than between two rough surfaces. The way that these differences in friction level are explained is by something called the "Coefficient of Friction." If the two surfaces slide against each other quite easily, the coefficient of friction is low. The lowest theoretical value is zero, but this is virtually impossible to reach. The upper level values are usually around one while the highest values reach, at most, a value of two.

Force of friction: $F_{fric} = \mu N$

Where
F_{fric} = the force of friction in newtons (N)
μ = the coefficient of friction
N = the normal force = $W\cos\theta$ in newtons (N)
Note that when the angle θ is zero degrees (0°), $N = W$

When you walk into a dry bathtub, your feet easily hold the tub. Once the tub gets wet, it feels a bit more slippery and your feet do not hold on so well. Then you spill the shampoo or some oil all over the bathtub, it is very difficult to remain standing on that surface. The dry friction between your feet and the dry bathtub has a moderately high coefficient of friction. The water reduces the coefficient of friction substantially, and the shampoo or oil almost eliminates it, bringing the coefficient of friction to near zero.

Static Friction

Static friction occurs between two stationary objects. A pair of shoes on the floor, tools on a workbench, art paint tubes and brushes on an easel, and boxes of birthday presents on a table all exhibit static friction when they are not moved. They "adhere" to the surface on which they stand. Remember that once you get a stationary object moving, it seems to move with less effort. The force of your push exceeded the force of static friction. It required a stronger push to get it moving initially and that is the effect of static friction. That heavy birthday present won't budge at first, but once you get it sliding, it is easy to keep it moving.

Force of static friction: $F_{fric} = \mu_s N$

Where
F_{fric} = the force of friction in newtons (N)
μ_s = the coefficient of static friction
N = the normal force = $W\cos\theta$ in newtons (N)

Note that when the angle θ is zero degrees (0°), $N = W$

Kinetic Friction

Kinetic friction is a continuous resistive force that acts between objects when they are moving against each other. Essentially, once the force applied to the object overcomes static friction, kinetic friction comes into play.

When we walk on ice, there is some friction between our boots and the ice that allows us to keep our traction and, therefore, our balance. However, we know we can sometimes slide on the ice without losing our balance and we stop after a few meters. There is some resistance between our shoes and the ice when we are sliding. This is kinetic friction.

Force of kinetic friction: $F_{fric} = \mu_k N$

Where
F_{fric} = the force of friction in newtons (N)
μ_k = the coefficient of kinetic friction
N = the normal force = $W\cos\theta$ in newtons (N)
Note that when the angle θ is zero degrees (0°), $N = W$

The coefficient of kinetic friction (μ_k) is always less than the coefficient of static friction (μ_s).

$\mu_k < \mu_s$

The table lists the static friction and kinetic friction between various pairs of surfaces.

Surfaces	μ_s	μ_k
Aluminum–Aluminum	1.05–1.35	1.4
Aluminum–Mild Steel	0.61	0.47
Cast Iron–Cast Iron	1.1	0.15
Copper–Cast Iron	1.05	0.29
Copper–Mild Steel	0.53	0.36
Glass–Glass	0.9–1.0	0.4
Glass–Nickel	0.78	0.56
Ice – Ice	0.05–0.5	0.02–0.09
Leather–Oak (parallel grains)	0.61	0.52
Nickel–Nickel	0.7–1.1	0.53
Oak–Oak (across grains)	0.54	0.32
Oak–Oak (parallel grains)	0.62	0.48
Steel–Zinc plated on Steel	0.5	0.45
Hard Steel – Hard Steel	0.78	0.42
Mild Steel–Brass	0.51	0.44
Mild Steel–Lead	0.95	0.95
Mild Steel–Mild Steel	0.74	0.57
Clean Wood – Clean Wood	0.25–0.5	0.17
Zinc–Cast Iron	0.85	0.21

Example 1: Static Friction

The weight of a block is 500 N. The block is on a flat surface with a coefficient of static friction of 0.7.

Using the formula for the force of static friction: $F_{fric} = \mu_s N$

The magnitude of the Normal force is the same as the weight, 500 N. The direction is opposite. The normal force is $N = -500$ N.

$F_{fric} = \mu_s N = (0.7)(-500 \text{ N}) = -350$ N.

Therefore, this block requires 350 N of force applied horizontally to get it to move from a static position.

Example 2: Friction

The weight of a block is 600 N. The block is on a surface at an angle of 30° with a coefficient of static friction of 0.7 and a coefficient of kinetic friction of 0.5.

Force of static friction: $F_{fric} = \mu_s N$

The magnitude of the normal force is the same as the weight, but opposite in sign, $N = -600$ N cosθ.

$$F_{fric} = \mu_s N = (0.7)(600 \text{ N } (\cos(30°)) = 420 \text{ N } (0.866) = 364 \text{ N}.$$

Therefore, this block requires 363.7 N of force applied at an angle of 30° down the ramp to get it to move from a static position at this angle.

Once the block is moving the kinetic friction is the resistive force.

$$F_{fric} = \mu_k N = (0.5)(600 \text{ N } \cos(30°)) = 300 \text{ N } (0.866) = 260 \text{ N}.$$

Therefore, this block requires 260 N of force applied at an angle of 30° down to keep it moving at this angle.

Dry Friction and Fluid Friction

Friction can also be segregated into types based on the conditions or types of materials involved in the interaction. This can be either dry friction or fluid friction. Dry friction exists between clean dry surfaces. If the conditions are clean and dry, the surfaces will exhibit both static and kinetic forms of friction. This is what exists when a dry clean box is on a dry clean table.

Fluid or viscous friction exists between surfaces when at least one surface is a liquid or gas. A good example is a water slide. We know that without the water, the skin of the body adheres to the plastic and we do not move. Therefore, there must be sufficient water flow so we do not receive friction burns while trying to have fun at a water park.

Other Effects of Friction

Friction causes resistance. This can result in wasted energy. Friction usually creates heat. Friction can also cause components to wear out when they are used for a long time. We know that if an engine is not maintained, or the amount of oil is not kept at the proper level or allowed to become very contaminated, the engine will not last as long. This is because the friction is increased, causing more heat, more wear on components and consequently even more heat and damage.

Without friction we could not drive, and could not even stand on the ground the way we do.

Concept Reinforcement

1. Explain the concept of friction.

2. Explain static friction.

3. Explain kinetic friction.

4. The weight of a block of nickel is 500 N. The block is on a flat surface of glass with a coefficient of static friction (0.78) and kinetic friction (0.56) as listed in the table. Determine the forces of static friction and kinetic friction.

5. The weight of a block of cast iron is 600 N. The block is on a surface at an angle of 30°, which is also cast iron. Using the coefficients of static friction (1.1) and kinetic friction (0.15) from the table, determine the forces of static friction and kinetic friction.

Section 3.2 – Structural Support Beams

Section Objective

- Describe beams and their uses in mechanical engineering

Introduction

This section presents the structure of a beam and the stresses that are typically experienced by beams. A beam is a long, narrow and stiff structural member that is used like the skeleton of a building. They are usually made from hard but flexible materials such as metals and wood. Vertical beams can be constructed of stone and concrete because both are stronger in compression than in tension or shear.

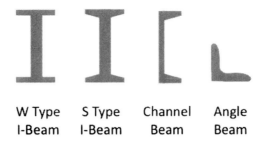

W Type   S Type   Channel   Angle
I-Beam   I-Beam   Beam   Beam

Description of a Beam

A beam is the term that refers to sections of a support structure. They are normally straight sections, but there are reasons that they may be other than straight.

Beams support a load. The strength of a beam relates to the amount the beam will bend when a particular load is placed on the beam.

The shape of a beam influences the ability of the beam to withstand a load and resist bending due to the applied load. A typical beam has a length that is many times larger than the dimension of the height or width. When discussing the various common shapes of a beam, this generally refers to the cross section.

The W Type I-Beam

The structure of a W Type I-Beam is a capital I, with both the top and bottom sections that are flat plates.

The S Type I-Beam

The structure of an S Type I-Beam is also a capital I, but the base and the cap are tapered towards the vertical section, providing additional stability.

The Channel Type Beam

The structure of a Channel Type Beam is like a vertical half of a capital I (half of the S Type I-Beam). The base and the cap are tapered toward the vertical section as in the S Type I-Beam.

The Angle Type Beam

The structure of an Angle Type Beam approximates a corner of a square with a horizontal base and a vertical portion, both of which are about half the height of the vertical section of a standard I-Beam.

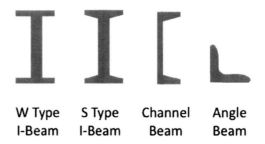

W Type   S Type   Channel   Angle
I-Beam   I-Beam   Beam   Beam

Stresses on Beams

Most beams carry vertical loads; i.e. loads from above as in the figure. In virtually all situations where beams are used, the objective is to transfer the load to vertical supports, such as columns, walls or other beams.

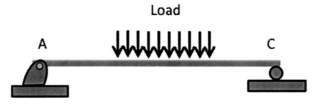

A load applied to a beam

Within construction applications, the material of beams is usually steel, while some applications require steel reinforced concrete, wood or aluminum. The type of beam used is based on the specific application for the beam.

Beams experience various forms of stress: compressive, tensile and shear stresses.

Compressive Stress

When the load is from the top, as in the figure, the upper side of the beam experiences compressive stress along the horizontal length of the beam. As can be seen in the second figure, an exaggerated amount of deflection occurs when the top bends down under the force of the load. This causes the horizontal compressive stress on the upper ½ portion of the beam.

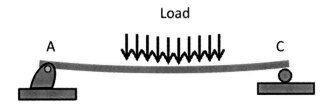

Load

A   C

A load applied to a beam with the beam deflecting as a result

Tensile Stress

On the other hand, the bottom side of the beam experiences tensile stress along the horizontal length of the beam. Under the force of the load, the bottom side of the beam extends out with the deflection. This is the cause of the tensile stress.

Shear Stress

Because the load is centered on the beam and the supports are at the ends of the beam, there is a strong possibility that shear stress exists within the beam itself. If we divide the beam up into many vertical layers, we can more readily explain the shear stresses.

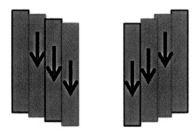

A load applied to a beam causes shear stress within the vertical portions of the beam.

Starting from the center and moving to the left, the central layers of the beam, where the load is applied, have relatively more force pushing the layers down compared to the next layer to the left.

The same concept applies as we move from the center to the right. We see the same situation, with the center portions experiencing more downward force than those layers to the right.

This is an important consideration as we move from theory (an assumption that the members of a structure, the beams, are weightless and do not deform) to a more realistic approach. All beams have mass and weight. They experience external loads, which cause compressive and tensile stresses within the beam. These loads also cause the beams to experience shear stresses.

Also, because the load is being supported by single-point structural supports, the beam experiences a moment. This is a bending moment, which is sometimes also called internal torque.

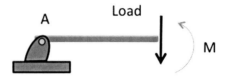

A portion of a beam in equilibrium under a load must also experience a bending moment in order to be in equilibrium.

Concept Reinforcement

1. Explain the description of a beam.

2. Name the forces experienced by a beam.

3. Explain where tensile and compressive stresses are experienced in a beam.

4. Describe shear stresses.

5. Describe bending moment.

Section 3.3 – Force and Moment

Section Objective

- Discuss axial force, shear force, and bending moment

Introduction

This section presents the concepts of axial force, shear force and bending moments. An axial force is experienced by the structural member parallel to the member. This can be either a tensile or compressive force. When an axial force is experienced by a specific cross-sectional area this produces an axial stress which can be either a tensile or a compressive stress. A shear force is what the material experiences when two layers slide past each other like two books sliding past each other. A bending moment is experienced by portions of a structural member that is in equilibrium when only a portion of the member is examined.

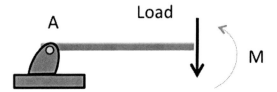

Within a fraction of a beam a bending moment will exist in equilibrium.

Axial force

Axial force is a force experienced by a structural member in the direction along the length of the member. Axial force is usually drawn as a tensile force, with the direction moving out of the member and normal to the cross section of the member. Axial force is the force experienced by an axial member, which is a member that experiences forces only at the ends.

Using this convention, with the force always pointing out of the axial end, if the value of the force is positive, it is then a tensile force and if it is negative, it is then a compressive force.

Axial Stress

Axial stress is measured in force per unit area. These measures include newtons per square meter (N/m^2) and pounds per square inch (psi). Therefore, axial stress is similar to the concept of pressure in a fluid. Both use the same base units of force per unit of cross sectional area.

Every material has a maximum axial stress that it can bear. When used in construction, the engineering design must limit the maximum load so that the axial stress threshold is not exceeded.

Axial Force

Axial Stress; $\sigma = F/A$ Area; $A = \pi r^2$

Axial force is given in newtons (or pounds). The area is given in square meters or (square inches).

Axial Stress: $\sigma = \dfrac{F}{A}$

Where:
σ = shear stress in newtons per square meter (N/m²)
F = force (load) in newtons (N)
A = area in square meters (m²)

Example 1: Axial Force

With a tensile force of 50 kN on an axial member, calculate the axial stress for a radius of 10 cm and 5 cm.

First, determine the cross sectional area of the member in square meters.

For a radius of 10 cm (0.10 m) the area is:

$$A = \pi r^2 = \pi (0.10 \text{ m})^2 = \pi (0.01 \text{ m}^2) = 3.14 \times 10^{-2} \text{ m}^2$$

The axial stress is: $\sigma = \dfrac{F}{A} = \dfrac{50 \text{ kN}}{3.14 \times 10^{-2} \text{ m}^2} = 1.6 \times 10^6 \text{ N/m}^2$

For a radius of 5cm (0.05m) the area is:

$$A = \pi r^2 = \pi (0.05 \text{ m})^2 = \pi (0.0025^2) = 7.85 \times 10^{-3} \text{ m}^2$$

The axial stress is: $\sigma = \dfrac{F}{A} = \dfrac{50 \text{ kN}}{7.85 \times 10^{-3} \text{ m}^2} = 6.37 \times 10^6 \text{ N/m}^2$

Note the axial stress increases by a factor of four when the radius is halved.

Shear Force (V)

Within the structure of a beam, the material can be considered to exist in layers. When there is a concentrated force (load) acting on a beam at a point, the material at that point will tend to move in the direction of the force. The material layers around this point will resist movement and hence there will be a force differential. The layers furthest away from the point force move least. This differential is called transverse shear stress and is always parallel to the direction of the acting force.

If we have a load centered on a beam and the supports are at the left and right ends then transverse shear stress exists within the material of the beam itself with more transverse shear stress close to the application point of the load. Starting from the center and moving to the left along the beam, the central portions of the beam have relatively more force pushing the layers down compared to the next layer to the left.

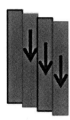 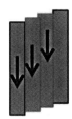

Shear stress occurs when a concentrated force acts on a portion of the material. The shear stress occurs between the layers of the material.

Again, moving from the center to the right, we see the same trend. The center layers experience more downward force than those layers to the right.

The transverse shear stress is typically determined at a specific distance from one of the ends. It is dependent upon the strength of the material that makes up the beam, the actual shear force applied at the location of interest, and the form of the beam, which then defines the moment of inertia of the beam, the load applied, as well as the moment of the area.

The typical construction is a vertical load applied to a horizontal beam.

In this case the formula for the shear stress (τ) is:

Shear Stress: $\tau = \dfrac{VQ}{It}$

Where:
τ = the shear stress in newtons/ square meters (N/m²) or pascals (Pa)
V = the shear force in newtons (N)
Q = the moment of the area (about the neutral axis) in cubic meters (m³)
I = the moment of inertia of the beam about the centroidal axis in meters to the fourth power (m⁴)
t = the width of the beam at the location the shear stress is measured in meters (m)

Bending moment

Through examining a portion of a beam in equilibrium we see the left support providing an upward force and the load being a downward force at the fractional end of the beam. It is assumed that the beam is supported on both ends and is in equilibrium; i.e. it is not moving. Here we are examining just a fraction of the beam because everywhere along the length of the beam the condition of equilibrium should also exist.

Within a fraction of a beam a bending moment will exist in equilibrium.

Under equilibrium, the external forces will sum to zero in both the vertical and horizontal directions. There are no horizontal forces in this instance. For the vertical forces to sum to zero, the force applied by the vertical support must be both equal and opposite to the load.

Also, for equilibrium to exist, we know that the moments about the left end must sum to zero. Therefore, an opposing moment must also be present at the fractional end of the beam, as shown in the figure. This is called a bending moment, which is sometimes also called internal torque. This bending moment is proportional to the distance from the left end and increases to a maximum at the point on the beam where the load is actually applied.

Shear Force and Bending Moment

When examining any point along the beam at equilibrium conditions, there will be a shear force and a bending moment acting on the beam at that position. This is required for the beam to be in equilibrium all along its length.

Concept Reinforcement

1. Explain the assumptions associated with an axial force.

2. Explain axial force and axial stress.

3. Explain mechanical equilibrium.

4. Define shear force.

5. Explain why everywhere along a beam in equilibrium there will be a shear force and a bending moment.

Section 3.4 – Shear Force and Bending Moment

Section Objective

- Draw and interpret shear force and bending moment diagrams

Introduction

This section presents the analytical concept of the shear force and bending moment. Beginning with axial members which experience forces only at the ends and non-axial members which experience forces anywhere along the member this section presents shear forces and bending moments that are part of the equilibrium conditions of a structural member. With any point load acting on a beam, different portions of a beam experience different amounts of load resulting in shear forces. Along the length of the beam the shear forces experienced change. To maintain equilibrium conditions along the length of the beam bending moments must exist. Along the length of the beam the bending moment also changes from one end to the other end of the beam.

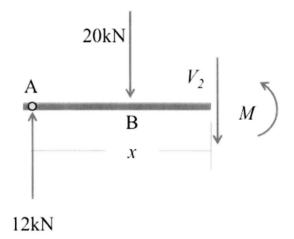

Free-body diagram of a portion of the Beam

Axial Members

An axial member is a structural section that experiences forces only at the ends. The result is either a tensile force or a compressive force experienced by the entire member. There is also the potential for a moment to be experienced by the member.

Axial members experience forces only parallel to the beam and only at the ends.

Axial members also experience stresses and deformations as a result of the forces parallel to the member itself.

Non-Axial Members

A non-axial member is a structural section that experiences forces anywhere along the length of the member. These forces can be experienced at any angle. The result is a more complicated set of issues related to the forces, stresses, moments and deformation of the member. To demonstrate these issues we use a standard horizontal beam carrying a load perpendicular to the beam.

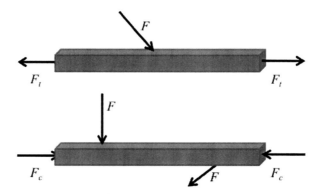

Non-Axial Members experience forces at any location
along the length of the beam, applied at any angle.

Shear forces

Shear forces are experienced by the beam under equilibrium conditions, due to the load acting on the material of the beam.

Bending moments

Bending moments (also called internal torque) will also be experienced under equilibrium conditions due to the combination of the load pushing down and the supports pushing up.

The shear forces and bending moments both need to be understood because they produce the stresses, both axial and shear, experienced by the beam. The shear force and bending moment are usually examined together primarily because a beam with a load and in equilibrium will experience both shear forces and bending moments at any position along the length of the beam.

Solving for Shear Force and Bending Moments

Here we examine how to quantify the shear forces and bending moments using an example. The steps that are followed to obtain the solution are the following:

Step 1: Examine the description of the problem.

Step 2: Create a free-body diagram for the problem.

Step 3: Draw the shear force and bending moment diagram

Step 4: Define the relationships that exist in formulas with the aim to have as many equations as there are unknown elements (forces and moments).

Demonstration Problem

The length from point A to point C is 20 m and from point A to point B is 8 m. The load is 20 kN.

Step 1: Description of the Problem

This description of the problem shows that the load is off center to the left which indicates that more of the load is supported by the left support. This additional force will be proportional to the fraction of the length of the beam.

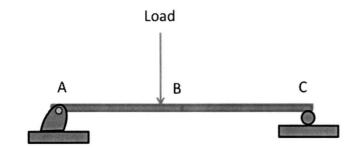

A Horizontal beam with a vertical load at point B experiences shear forces and bending moments.

Step 2: Free-Body Diagram

To create the free-body diagram we know that under equilibrium conditions the vertical load is countered by an equal amount of force in the opposite direction which is the sum of the upward forces from the two supports. This assumes that the beam itself does not deform and only transfers the forces to the supports.

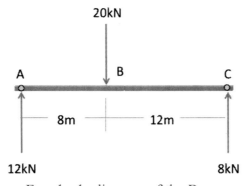

Free-body diagram of the Beam

If the load was in the center of the beam each support would carry 10 kN but since the load is off center we can solve the actual load using fractions.

Support Force on point A: $F_A = 20 \text{ kN}\left(\dfrac{12 \text{ m}}{20 \text{ m}}\right) = 12 \text{ kN}$

Support Force on point C: $F_C = 20 \text{ kN}\left(\dfrac{8 \text{ m}}{20 \text{ m}}\right) = 8 \text{ kN}$

Step 3: Shear Forces and Bending Moments

To understand the shear forces and the bending moments we examine a portion of the beam to the left of the load.

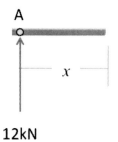

12kN

Free-body diagram of a portion of the beam to the left of the load

The entire beam is in equilibrium and therefore each section of the beam is in equilibrium. This is the basis of the determination of the shear forces and the bending moments.

Step 4: Define the equations

The beam at a distance "x" from point A experiences a shear force (V) and a bending moment (M).

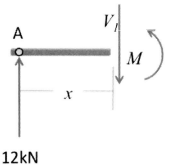

12kN

Free-body diagram of a portion of the beam to the left of the load

To determine the forces, we again turn to the fact that the vertical forces are in equilibrium and sum to zero. Therefore, the portion of the vertical load is equal and opposite the force upward that the beam experiences from the left support.

Hence, the portion of the load is 12kN; i.e. $V_1 = 12 \text{ kN}$ with the direction downward.

The bending moment that the beam experiences is determined by the fact that the moments about the left end must also sum to zero. Using clockwise as positive, we have:

Sum of moments: $V_1x - M = 0$ and therefore $M = V_1x$

Including that $V_1 = 12$ kN we have: $M = (12$ kN$)x$

For example, if the distance $x = 6$ m the solution is $M = (12$ kN$)6$ m $= 72$ kNm

Now we examine a section of the beam on the other side of the load.

Here the calculation of the shear force includes the entire load as well as the support force at point A.

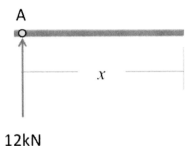

12kN

Free-body diagram of a portion of the beam to the right of the load

The equilibrium equation for the forces is:

12 kN $-$ 20 kN $- V_2 = 0$; i.e. $V_2 = (12\text{-}20)$ kN $= 8$ kN

Therefore the shear force (V_2) is 8 kN upward.

The bending moment that the beam experiences is again determined by the fact that the moments about the left end at point A must also sum to zero. Using clockwise as positive we again have the equilibrium equation for the moments:

Sum of moments: $(20$ kN$)(8$ m$)+V_2x - M = 0$ and therefore $M = 160$ kNm $+ V_2x$

Including that $V_2 = 8$ kN we have: $M = 160$ kNm $+ (8$ kN$)x = 160$ kNm 8 kN (x)

For example, if the distance $x = 15$m the solution is $M = 160$ kNm $+ (8$ kN$)(15$ m$) = 160$ kNm 120 kNm $= 40$ kNm.

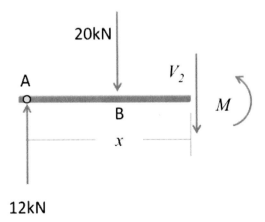

12kN

Free-body diagram of a portion of the beam

This infers that the direction shown for the moment (counterclockwise) is correct and the moment is 40 kNm in the counterclockwise direction.

From this data the Shear Force and Bending Moment Diagram is constructed

Shear Force and Bending Moment Diagram

Section 1: $0 < x < 8$ m
$V_1 = 12$ kN
$M_1 = 12$ kNx

Section 2: $8 < x < 20$ m
$V_2 = -8$ kN
$M_2 = 160$ kNm -8kNx

Shear force and bending moment equations describing the two sections

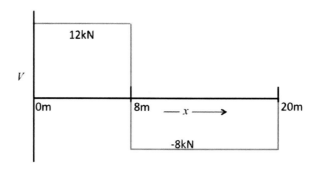

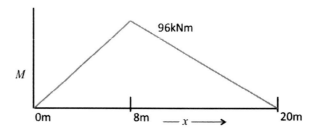

Shear force diagram (above) and bending moment diagram (below)

The shear force (V) is constant at 12 kN from the left support to the load and is constant at 8 kN from the load to the right support.

The bending moment increases linearly from zero (0) at the left support to a maximum of 96 kNm at the location of the load (at 8 m) and decreases linearly from the location of the load to the right support.

Concept Reinforcement

1. Explain the assumptions around an axial member.

2. Explain the assumptions around a non-axial member.

3. Explain shear force.

4. Define the bending moment.

Section 3.5 – Distributed Load Moment

Section Objective

- Explain distributed loads, bending moments, and shear forces

Introduction

This section presents the shear force and bending moment diagram of a distributed load which is a graphic display of the effects of a distributed load on a beam. This analysis is usually begun from the left-most joint of the beam. The shear force experienced by the beam begins at a maximum positive magnitude and gradually reduces to a maximum negative magnitude. The bending moment begins at zero at the left end, increases to a maximum at the centroid of the distributed load and then reduces again to zero at the right most end.

Distributed Load

Structural members may experience concentrated loads at a point or loads distributed over a portion or over the full length of the member. The typical free-body diagram shows the force experienced at a point. A distributed load appears differently in the free-body diagram.

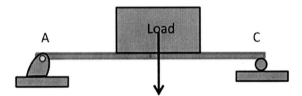

Typical free-body diagram with the load presented as acting from the center of the load

A distributed load is actually much more common than point loads and can be visualized as a block with mass resting on a structural member. The most common distributed load is the uniform load where a load of constant magnitude (per unit of length) is applied over the entire length of the beam.

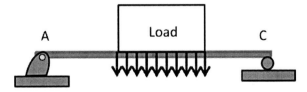

Distributed load presented as acting across the entire width of the load

The mass of the block experiences the gravitational acceleration and this creates the weight of the object or load. This load is spread evenly along the beam for the full length of the block.

The free-body diagram for a distributed load no longer shows a single force acting from the center of the object. This can complicate the equilibrium calculations but there is a common procedure to return us to a force applied at a point. That is called resolving the distributed loads and entails finding the centroid of the load (the center of gravity) and determine the total force applied by the load.

Resolving Distributed Loads

Starting with uniformly distributed loads the centroid is easily found. For a rectangle or a uniform load "area" the centroid is the intersection of the two diagonal lines.

Load

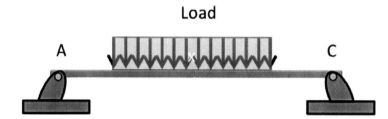

Distributed load presented as acting across the entire width of the load

Example 1: Resolving Distributed Loads

For a load that is 10 kN/m spread over a length of 3 meters find the centroid and the total load.

Load

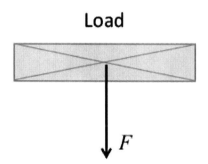

Finding the centroid of a uniformly distributed load using the intersection of diagonal lines. This is also a position at ½ the length and ½ the height.

The important dimension is the horizontal location of the force. This being a uniformly distributed load the resolved force will be applied at 1.5 meters from the end of the "block" of the load. Exactly where this is relative to a support structure must be determined when that data is available.

The total load is the force per unit length times the length. This is 10 kN/m × 3 m = 30 kN.

Therefore the resolved load will be a force of 30 kN applied at 1.5 m from the end of the section of the beam where the distributed load begins.

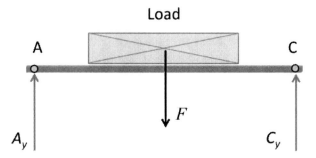

Load

The free-body diagram for a uniformly distributed load with the resolved force

Moving on to more complex loads it is also possible that the load is a uniformly increasing load. This load "area" could either appear like a triangle or like a triangle placed on top of a rectangle.

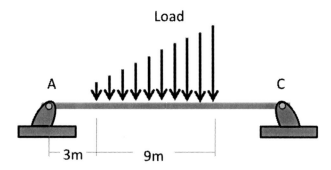

Load

The free-body diagram for a triangular distributed load

The centroid of a rectangle is found as described above. The centroid of a triangle is also a simple formula of the intersection of two lines, each at 1/3 from the wider part of the triangle, with one line parallel to one of the bases of the triangle.

Example 2: Triangular distributed load

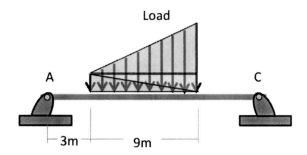

Load

The free-body diagram for a triangular distributed load
divided into three triangular shapes

Here the load begins on the left at 3 kN/m and ends on the right at 12 kN/m.

This can be divided up into three triangular distributions to resolve the force. This will result in three separate concentrated forces but this is for illustration of the process.

Triangle 1:

The base is 9 m and therefore the centroid is at 3 m from the left. The height is 3 kN/m at the left and zero at the right. The area of a right triangle is ½ *bh*.

$$F_1 = A_1 = \frac{1}{2}bh = \frac{1}{2}(9 \text{ m})(3 \text{ kN/m}) = \frac{27}{2} \text{ kN}$$

Triangle 2:

The base is again 9 m and therefore the centroid is at 3 m from the right. The height is 3 kN/m at the right and zero at the left. The area of a right triangle is ½ *bh*.

$$F_2 = A_2 = \frac{1}{2}bh = \frac{1}{2}(9 \text{ m})(3 \text{ kN/m}) = \frac{27}{2} \text{ kN}$$

Triangle 3:

The base is again 9 m and therefore the centroid is at 3 m from the right. The height is (1.2 kN/m – 3 kN/m) = 9 kN/m at the right and zero at the left. The area of a right triangle is ½ *bh*.

$$F_3 = A_3 = \frac{1}{2}bh = \frac{1}{2}(9 \text{ m})(9 \text{ kN/m}) = \frac{81}{2} \text{ kN}$$

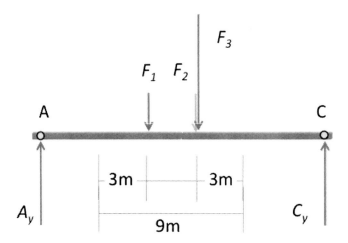

The free-body diagram for a triangular distributed load
divided into three triangular shapes

Bending Moment of a Distributed Load

The bending moment is the rotational expression of the force applied at a distance from a point. When the load is concentrated at a point the moment is calculated as the separation distance between the point and the load times the load.

Bending Moment: $M = Fd$

Where:
M = the bending moment in newton-meters (Nm)
F = the force (or load) in newtons (N)
d = the separation distance in meters (m)

When the load is uniformly distributed, the bending moment is different at different separation distances from the point to part of the load.

Let's examine how to quantify the shear forces and bending moments using an example with a uniformly distributed load.

The steps followed to obtain the solution are:

Step 1: Examine the description of the problem.

Step 2: Create a Free-Body Diagram for the problem.

Step 3: Draw the Shear Force and Bending Moment Diagram.

Step 4: Define the relationships that exist in formulas with the aim to have as many equations as there are unknown elements (forces and moments).

Shear Force and Bending Moment Diagram for Distributed Loads

Step 1: Description of the Problem

This description of the problem shows that the uniformly distributed load is off center to the left, which indicates that more of the load is supported by the left support. This additional force will be proportional to the fraction of the length of the beam.

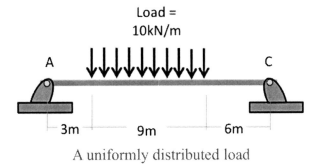

A uniformly distributed load

Step 2: Free-Body Diagram

To create the free-body diagram we know that under equilibrium conditions the vertical load is countered by an equal amount of force in the opposite direction, which is the sum of the upward forces from the two supports. This assumes that the beam itself does not deform and only transfers the forces to the supports.

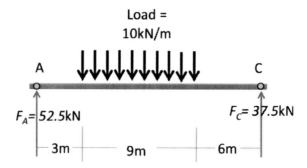

Load =
10kN/m

A   C

$F_A = 52.5kN$ $F_C = 37.5kN$

3m   9m   6m

Free-body diagram for a uniformly distributed load

If the load was concentrated in the center of the beam, each support would carry 45 kN. However, since the load is off center, we can solve the actual load using fractions. We first resolve the load to a concentrated load at the centroid. The magnitude of the load is 10 kN/m × 9 m = 90 kN. The centroid is located at ½ the length of the load = 4.5 m from the end of the load.

From the left side, this is 3 m + 4.5 m = 7.5 m.

From the right side, this is 6 m + 4.5 m = 10.5 m.

Support Force on point A: $F_A = 90 \text{ kN} \left(\dfrac{10.5 \text{ m}}{18 \text{ m}} \right) = 52.5 \text{ kN}$

Support Force on point C: $F_C = 90 \text{ kN} \left(\dfrac{7.5 \text{ m}}{18 \text{ m}} \right) = 37.5 \text{ kN}$

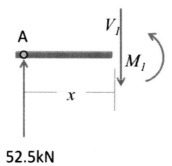

A V_1

M_1

x

52.5kN

Free-body diagram of a portion of the beam to the left of the load

Step 3: Shear Forces and Bending Moments

To understand the shear forces and the bending moments, we examine a portion of the beam to the left of the load.

The entire beam is in equilibrium and therefore each section of the beam is in equilibrium. This is the basis of the determination of the shear forces and the bending moments.

Step 4: Define the equations

The beam, at a distance "x" from point A, experiences a shear force (V) and a bending moment (M).

To determine the forces, we again turn to the fact that the vertical forces are in equilibrium and therefore sum to zero. Therefore the portion of the vertical load is equal and opposite the force upward that the beam experiences from the left support.

Hence, the portion of the load is 52.5 kN; i.e. $V_1 = 52.5$ kN with the direction downward.

The bending moment that the beam experiences is determined by the fact that the moments about the left end must also sum to zero. Using clockwise as positive we have:

Sum of moments: $V_1 x - M_1 = 0$ and therefore $M_1 = V_1 x$

Including that $V_1 = 52.5$ kN we have: $M_1 = (52.5\ \text{kN})x$

For example, if the distance $x = 3$ m the solution is $M_1 = (52.5\text{kN})3\ \text{m} = 157.5$ Nm

Now we examine a section of the beam on the other side of the load.

Here the calculation of the shear force includes the entire load as well as the support force at point A.

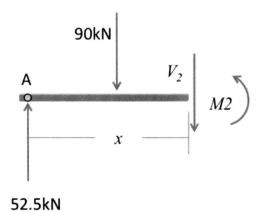

Free-body diagram of a portion of the beam

The equilibrium equation for the forces is:

52.5 kN – 90 kN – V_2 =0; i.e. $V_2 = (52.5 - 90)$ kN = 37.5 kN

Therefore, the shear force (V_2) is 37.5 kN upward.

The bending moment that the beam experiences is again determined by the fact that the moments about the left end at point A must also sum to zero. Using clockwise as positive we again have the equilibrium equation for the moments:

Sum of moments: $(90\ \text{kN})(7.5\ \text{m}) + V_2 x - M = 0$ and therefore $M = 675$ kNm + $V_2 x$

Including that $V_2 = 37.5$ kN we have: M = 675 kNm + $(37.5\ \text{kN})x = 675$ kNm 37.5 kN(x)

For example, if the distance $x = 13$ m the solution is M = 675 kNm + $(37.5\ \text{kN})(13\ \text{m}) = $ 675 kNm 488 kNm = 188 kNm.

This infers that the direction shown for the moment (counterclockwise) is correct and the moment is 188 kNm in the counterclockwise direction.

Now we examine a section of the beam where the uniformly distributed load is applied.

In this case, the calculation of the shear force includes a portion of the entire load, as well as the support force at point A.

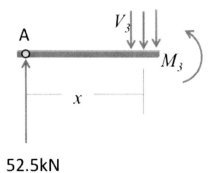

52.5kN

Free-body diagram of a portion of the beam in the uniformly distributed load

The equilibrium equation for the forces is:

$$52.5 \text{ kN} - (10 \text{ kN/m}) (x - 3 \text{ m}) - V_3 = 0$$

$$V_3 = 52.5 \text{ kN} + 10\text{kN/m} (3 \text{ m}) - (10 \text{ kN/m})x$$

$$V_3 = 82.5 \text{ kN} - (10 \text{ kN/m})x$$

Therefore the shear force (V_3) is 82.5 kN – (10 kN/m)x with the direction dependent upon the sign of the value. If the shear force is positive then the shear force is downward.

x-range	V – Shear Force
0 to 3 m	52.5 kN
3 to 12 m	52.5 kN to –37.5 kN
12 to 18 m	–37.5 kN

The bending moment that the beam experiences is again determined by the fact that the moments about the left end at point A must also sum to zero. Using clockwise as positive, we again have the equilibrium equation for the moments:

Sum of moments: $V_3x + 10 \text{ kN/m}(x - 3 \text{ m})(3+(x - 3)/2) - M_3 = 0$

Term 1: The shear force times the distance from Point A

Term 2: The fractional load applied at the middle point of the load, times the distance to point A

Solving for the moment M_3:

$$M_3 = V_3 x + 10 \text{ kN/m}(x-3 \text{ m})\left(3 \text{ m} + \frac{x-3}{2} \text{ m}\right) = V_3 x + (10 \text{ kN/m} - 30 \text{ kN/m})\left(3 \text{ m} + \frac{x-3}{2} \text{ m}\right)$$

The formula for the moment can be expanded using $V_3 = 82.5 \text{ kN} - (10 \text{ kN/m})x$

$$M_3 = \left(82.5 \text{ kN} - (10 \text{ kN/m})x\right)x + \left(10 \text{ kN/m}(x) - 30 \text{ kN/m}\right)\left(3 \text{ m} + \frac{x-3}{2} \text{ m}\right)$$

$$M_3 = 82.5 \text{ kN} - (10 \text{ kN/m})x^2 + \left(10 \text{ kN/m}(x) - 30 \text{ kN/m}\right)\left(3 \text{ m} + \frac{x-3}{2} \text{ m}\right)$$

For example, if the distance $x = 10$ m the solution is

$$M_3 = 82.5 \text{ kN}(10 \text{ m}) - (10 \text{ kN/m})(10 \text{ m})^2 + \left(10 \text{ kN/m}(10 \text{ m}) - 30 \text{ kN/m}\right)\left(3 \text{ m} + \frac{10-3}{2} \text{ m}\right)$$

$M_3 = 825 \text{ kNm} - (1{,}000 \text{ kNm}) + (100 \text{ kNm} - 30 \text{ kNm}) (6.5 \text{ m})$

$M_3 = 280 \text{ kNm}$

This inplies that at 10 m the direction shown for the moment (counterclockwise) is correct and the moment is 280 kNm in the counterclockwise direction.

From this data, the Shear Force and Bending Moment Diagram is constructed.

Shear Force and Bending Moment Diagram

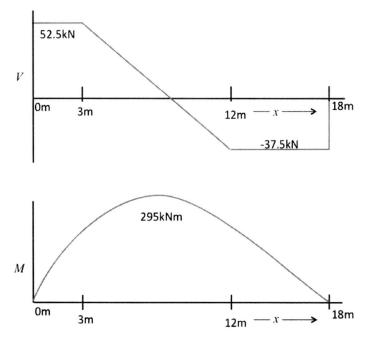

Shear Force Diagram (above) and Bending Moment Diagram (below)

Section 1: $0 \text{ m} \leq x < 3 \text{ m}$

The shear force (V_1) is constant at 52.5 kN from the left support to the start of the distributed load at $x = 3$ m.

The bending moment (M_1) increases linearly from zero (0) at the left support to a 157.5 kNm at $x = 3$ m.

Section 3: $3 \text{ m} \leq x \leq 12 \text{ m}$

The shear force (V_3) is constantly reducing by 10 kN/m. Starting at 52.5 kN at $x = 3$ m, V_3 reduces to 37.5 kN at $x = 12$ m.

The bending moment (M_3) continues to increase but not linearly. From $x = 3$ m to $x = 8$ m the moment increase to a maximum of 295 kNm. After that point the moment reduces, again not linearly, to 225 kNm at $x = 12$ m.

Section 2: $12 \text{ m} < x \leq 18 \text{ m}$

The shear force (V_2) is constant at 37.5 kN from the point at $x = 12$ to the right support at $x = 18$ m.

The bending moment (M_2) decreases linearly from 225 kNm at $x = 12$ to zero (0) at the right support $x = 18$ m.

The shear force (V_1) is constant at 52.5 kN from the left support to the start of the distributed load at $x = 3$ m.

Concept Reinforcement

1. Explain a distributed load.

2. Explain how to resolve a uniformly distributed load into a concentrated load.

3. Explain how to resolve a triangular force distribution into a concentrated load.

4. Explain the formula for the shear force in the section of beam with a distributed load.

5. Explain the formula for the bending moment in the section of beam with a distributed load.

Section 3.6 – Cables

Section Objective

- Describe cables and their uses in mechanical engineering

Introduction

This section presents the description of cables, both support cables and control cables. Support cables are usually seen in bridges and tents and work primarily in tension. Because of their flexibility, they cannot transmit compressive forces or bending moments. There are various ways that these cables are constructed all of which focus on maximizing the tensile strength. Control cables are also used primarily in tension and are often constructed with a sheath for protection, while the cable inside moves back and forth transmitting tensile forces.

Cables

A cable is a flexible structural support. In general, a cable is assumed to offer no resistance to bending. Therefore, a cable cannot transfer bending moments. In addition, because of this assumption, cables cannot transfer transverse loads and all loads are assumed to be represented as tensile forces along the cable (normal to the cross sectional area of the cable).

Cables do not have their own shape because they adapt to the loads that are placed on them. It is also assumed that the length of the cable material does not change its length as a result of the load it bears. This characteristic of a cable is called infinite axial stiffness.

Cables can be constructed using several parallel wires braided together along the length of the wire. They can also be constructed with a solid core with many parallel wires wound around the core. The primary objective of this construction is to create a structural element that is more flexible than a beam while also having the tensile strength comparable to the other structural members. As the amount of individual wires increases, the tensile strength of the group of wires which are components of the cable increases.

Parallel Wire Cable

Structural Cables

Primarily used as a bridge support, structural cabling is more flexible than solid structural members. It does, however, have the same material properties in terms of the amount of stress it can withstand.

Galvanized Bridge Wire

This is the basic wire used to produce parallel wire bridge cables. Various sizes are available, with a typical diameter of approximately 0.2 cm.

Galvanized Bridge Strand

A bridge strand is constructed of many bridge wires twisted together to form a single support. The bridge wires may have many different diameters. The bridge strand may consist of a few or many individual bridge wires.

Galvanized Bridge Rope

A bridge rope is constructed of several, typically six, bridge strands. These strands are usually braided around a center core strand.

Parallel Wire Cables

This is a cable that consists of many individual smaller wires, all of which are parallel to each other. These wires are not twisted or braided. They are usually delivered on huge reels and "spun" on to the bridge support. Once the planned amount of wire is spun out the wire is compacted together into a round cross section.

Closed Parallel Strand Cables

This consists of several parallel galvanized strands held tightly together and wrapped along the entire length by wire for protection.

Open Parallel Strand Cables

These cables consist of several parallel galvanized strands held close, but at a slight distance from each other, usually forming a rectangle. The rectangular matrix of these strands is bound at intervals by structural elements, which are riveted to the cables.

Cables are also used as a method to transmit either tensile force or compressive force through a space with obstruction that requires a flexible structural element. The advantage of a cable is that it is usually lighter weight and much less bulky than a mechanical combination of solid members designed to do the same thing.

Bicycles use various control cables

Types of Force Transmission

A cable system often includes the metal braided cable surrounded by a protective, as well as controlling, sheath or housing.

Linear Transmission (A Bowden cable)

Though the cable will have one or more turns along the path, the cable sheath will be stiffer in order to improve the transmission of a linear force. This can be a pushing force, which translates to compression on the cable from one end, transmitting a push to an element at the other end of the cable.

This can also be a pulling force, which translates to tension on the cable from the operator end. This transmits a pull to the other end of the cable.

Rotational Transmission

In this type of transmission, the cable is moved linearly into and out of a housing that translates this linear motion into a rotation of a wheel within the housing.

Angular Motion

When there is a need for a structural member to be able to flex in position and also transmit a turning or twisting motion, the choice is often a type of cable that remains and includes several "elbow" joints so that the angular motion is possible.

Control cable construction

Control cables can be constructed using various designs for wrapping the component wires. Each structure has its advantages for particular applications.

Wire rope cables are a braided set of narrow gauge wire with no external wrapping around the braid. These are extremely flexible, which results in a very low ability to transmit tensile forces and no ability to transmit compressive forces. These typically transmit force with low efficiency.

Helix cables consist of a straight core of wires wrapped by one or more layers of heavier wire. These are very flexible, but this flexibility results in a reduction in the amount of tensile or compressive forces that can be transmitted. Within the range of loads that are allowed, these cables are very efficient.

Flat wrap cables also have a core of wires, but the wrapping consists of one or more flat metal layers. These outer layers provide much stronger structural integrity and, as a result, flat wrap cables are able to transmit higher tensile and compressive loads even though these flat wrap cables are much less flexible.

Concept Reinforcement

1. Explain the purpose of a structural bridge cable.

2. Explain the construction of a structural bridge cable.

3. Explain the three levels of bridge cabling.

4. Explain the types of force transmission with control cables.

5. Describe the three types of control cable construction.

Section 3.7 – Loads Along Straight Lines

Section Objective

- Explain loads along straight lines

Introduction

This section presents the concept of the effects of a uniform load on the tension in a suspension cable. Generally, this uniform load causes the line to take the shape of a portion of a parabola. The formula of that parabola is given and a formula is presented that describes the tension in the cable relative to the uniformly distributed load, the minimum tension in the cable and the distance along the cable from the end.

Loads along straight lines

The forces and shape of a cable can be analyzed for several situations:

1. when the cable is carrying a uniformly distributed load, as with a suspension bridge.

2. when the cable itself is analyzed based on its deadweight load and no external load.

3. when the cable is carrying several non-uniform loads.

This section analyzes the situation when the cable is carrying uniformly distributed load.

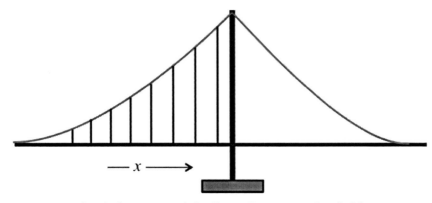

A load along a straight line of a suspension bridge

The concept of a load acting along a straight line is most clearly demonstrated by the top cable of a suspension bridge. The top cable runs at an angle from the top of the vertical support in the middle of the span. It gradually lowers until it reaches the base supports at the level of the bridge itself at one of the ends of the bridge.

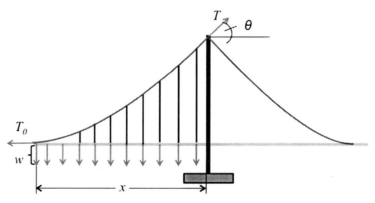

A load along a straight line of a suspension bridge

All along this cable, and at consistent vertical spacing, are vertical cables. These vertical cables transmit the load of the bridge uniformly to the cable through the entire length of the cable.

Each cable transmits a uniform fraction of the load to the cable. The total load is the weight of the bridge itself. As can be seen in the figure of the free-body diagram, the load (w) is the same at each vertical support cable. This uniform load, with an assumption of a massless cable (the mass of the cable does not enter into the equilibrium equations), creates a parabolic shape in the cable.

This situation can be described using the following:

At the left end of the cable the load on the cable is T_0. This is the minimum load that the cable experiences. At the right end of the cable, the load on the cable is T.

In the horizontal equation for equilibrium of the forces: $T_0 = T\cos\theta$.

In the vertical equation for equilibrium of the forces: $wx = T\sin\theta$.

Comparing the two equations, we have: $\dfrac{wx}{T_0} = \dfrac{T\sin\theta}{T\cos\theta} = \tan\theta$.

The tangent of a line is the slope. To define the shape of the cable itself, this value for the slope is used in an equation of a parabola.

The shape of the cable is: $y = \dfrac{1}{2}\dfrac{w}{T_0}x^2$.

Where:
y = the vertical position of the cable from the ground in meters (m)
w = the load per unit length in newtons per meter (N/m)
T_0 = the minimum tension on the cable in newtons (N)
x = the horizontal distance from the left end in meters (m)

To determine the tension in the cable, both of the equations $T_0 = T\cos\theta$ and $wx = T\sin\theta$ are squared and summed:

The square of $T_0 = T\cos\theta$ is: $T_0^2 = T^2\cos^2\theta$

The square of: $wx = T\sin\theta$ is: $w^2x^2 = T^2\sin^2\theta$

The sum of the two equations is: $T_0^2 + w^2x^2 = T^2\cos^2\theta + T^2\sin^2\theta$, which can be reduced to $T_0^2 + w^2x^2 = T^2(\cos^2\theta + \sin^2\theta)$. Using the identity $\sin^2\theta + \cos^2\theta = 1$, this equation simplifies to: $T_0^2 + w^2x^2 = T^2$ or $T^2 = T_0^2 + w^2x^2$.

This can be rearranged to: $T^2 = T_0^2\left(1 + \dfrac{w^2}{T_0^2}x^2\right)$

Taking the square root of both sides gives: $T = T_0\sqrt{1 + \dfrac{w^2}{T_0^2}x^2}$. This presents a formula that

shows that the tension T is equal to the minimum tension T_0 when $x = 0$.

Example 1: Load along a straight line

A bridge with a uniform load (w) of 100 MN/m (100×10^6 newtons/meter) is suspended by two cables. For a section of the bridge that is 100 m long, define the equations that describe the tension in the cable and the shape of the cable.

First, the total load is the uniform load (w) measured over the length (x).

$wx = 100$ MN/m$(100$ m$) = 10.0 \times 10^9$ N

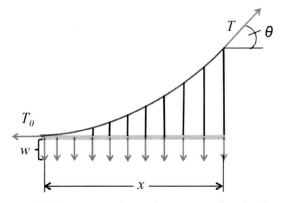

A load along a section of a suspension bridge

To determine the value of the tension T requires an angle θ and use of the equation:

$T \sin\theta = wx$ which can be solved for tension T:

Tension in a span of a cable: $T = \dfrac{wx}{\sin\theta}$.

For example, if the angle is 30° the sin30 is 0.5.

Therefore the tension is: $T = \dfrac{wx}{\sin\theta} = \dfrac{10 \times 10^9 \text{ N}}{0.5} = 20 \times 10^9 \text{ N}$

Of course, this is the tension of the two cables so each cable would have ½ of this value.

Using the equation describing the tension, we can assume the angle of 30° and find the minimum tension T_0 where $T_0 = T\cos\theta$.

$$T_0 = T\cos\theta = 10 \times 10^9 \text{ N} \cos 30° = 10 \times 10^9 \text{ N} (0.866) = 8.66 \times 10^9 \text{ N}$$

This is the tension in a single cable (hence the value of T in the equation is ½ that shown above).

The equation describing the tension is: $T = T_0 \sqrt{1 + \dfrac{w^2}{T_0^2} x^2}$.

Adding the values from above:

The equation describing the tension is: $T = 8.66 \times 10^9 \text{ N} \sqrt{1 + \dfrac{100 \times 10^6 \text{ N/m}}{8.66 \times 10^9 \text{ N}} x^2}$

The shape of the cable is: $y = \dfrac{1}{2} \dfrac{w}{T_0} x^2$.

Adding the values from above (using ½ of the load for each cable):

$$y = \dfrac{1}{2} \left(\dfrac{50 \times 10^6 \text{ N/m}}{8.66 \times 10^9 \text{ N}} \right) x^2 = \dfrac{1}{2} \left(5.77 \times 10^{-3} /\text{m} \right) x^2$$

The shape of the cable is: $y = (2.9 \times 10^{-3}/\text{m}) x^2$.

Concept Reinforcement

1. Explain the concept of a load along a straight line.

2. Explain a typical bridge structure with a long straight line load.

3. Explain the equation for the tension in the cable.

4. Define the equation for the shape of the cable.

Section 3.8 – Loads along Cables

Section Objectives

- Explain loads along cables

Introduction

This section presents the concept of the effects of the deadweight of the cable itself on the shape of the cable and the tension within the cable. The shape is slightly different from that expected for a uniformly distributed load like a bridge. The deadweight of the cable causes the cable to take the shape called a catenary, which is similar to a parabola. This describes the shape that a cable hanging freely between two supports obtains. The tension in the cable caused by the deadweight of the cable itself also has a slightly different result because the formula includes the length along the cable rather than the horizontal length as in the uniformly distributed load.

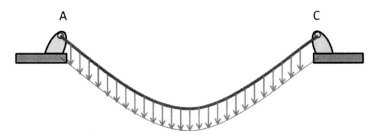

A cable suspended between two supports

Loads along Cables

The forces and shape of a cable can be analyzed for several situations; A. when the cable is carrying a uniformly distributed load, as with a suspension bridge, B. when the cable itself is analyzed based on its deadweight load and no external load and C. when the cable is carrying several non-uniform loads. This section analyzes the situation when the cable itself is analyzed based on its deadweight load and no external load.

A cable suspended between two supports

When talking about loads along cables, this refers to the deadweight load of the cable itself with no external forces or loads included in the analysis.

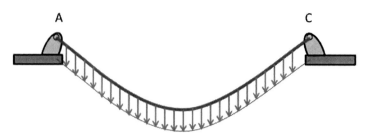

A cable suspended between two supports

The shape of a cable that is subjected to a uniform load, such as a cable of a suspension bridge, is a parabola.

When examining just the deadweight load of the cable itself, the shape of the cable is similar to a parabola called a catenary, but requires a different formula. The word catenary comes from a term used to describe the shape of a chain or rope hanging between two supports.

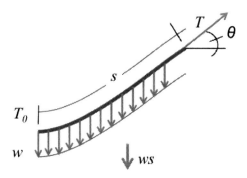

The free-body diagram of a section of the deadweight load along a cable

This situation can be described using the following:

At the left end of the cable, the load on the cable is T_0. This is the minimum load that the cable experiences. At the right end of the cable, the load on the cable is T.

In the horizontal equation for equilibrium of the forces: $T_0 = T\cos\theta$.

In the vertical equation for equilibrium of the forces: $ws = T\sin\theta$.

Note that here the distance (s) is along the cable and not in the x-direction.

Comparing the two equations, we have: $\dfrac{w}{T_0}s = \dfrac{T\sin\theta}{T\cos\theta} = \tan\theta$

The tangent of a line is the slope. Using this concept and the Pythagorean theorem as $s^2 = x^2 + y^2$ plus some advanced derivation, we can develop the formula for the catenary shape of the cable.

Initially: $y = \dfrac{T_0}{2w}\left(e^{\frac{wx}{t_0}} + e^{-\frac{wx}{t_0}} - 2\right)$

Through the identity for the hyperbolic cosine function: $\cosh = \dfrac{e^a - e^{-a}}{2}$, this formula for the shape of the cable can be simplified.

The shape of the cable is: $y = \dfrac{T_0}{w}\left(\cosh\left(\dfrac{w}{T_0}x\right) - 1\right)$

Where:
y = the vertical position of the cable from the ground in meters (m)
w = the load per unit length in newtons per meter (N/m)
T_0 = the minimum tension on the cable in newtons (N)
x = the horizontal distance from the left end in meters (m)
$\cosh$ = the hyperbolic cosine

Tension in the Cable

To determine the tension in the cable, both of the equations $T_0 = T\cos\theta$ and $ws = T\sin\theta$

are squared and summed:

The square of $T_0 = T\cos\theta$ is: $T_0^2 = T^2\cos^2\theta$

The square of: $ws = T\sin\theta$ is: $w^2s^2 = T^2\sin^2\theta$

The sum of the two equations is: $T_0^2 + w^2s^2 = T^2\cos^2\theta + T^2\sin^2\theta$, which can be reduced to $T_0^2 + w^2s^2 = T^2(\cos^2\theta + \sin^2\theta)$. Using the identity $\sin^2\theta + \cos^2\theta = 1$, this equation simplifies to: $T_0^2 + w^2s^2 = T^2$ or $T^2 = T_0^2 + w^2s^2$.

This can be rearranged to: $T^2 = T_0^2\left(1 + \dfrac{w^2}{T_0^2}s^2\right)$

Taking the square root of both sides gives: $T = T_0\sqrt{1 + \dfrac{w^2}{T_0^2}s^2}$

This presents a formula that show the tension T is equal to the minimum tension T_0 when $s = 0$.

Example 1: Load along a Cable

A cable with a deadweight load (w) of 40 N/m is suspended between two supports. For a section of the cable (s) that is 50 m long, define the equations that describe the tension in the cable and the shape of the cable. Assume the angle θ is 30°.

First, the total load is the uniform load (w) measured over the length (s).

$ws = 40 \text{ N/m}(50 \text{ m}) = 2{,}000 \text{ N}$

Determining the value of the tension T requires an angle θ and use of the equation:

$T\sin\theta = ws$, which can be solved for tension T:

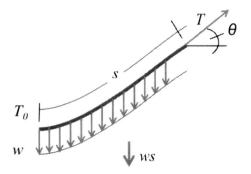

A load along a cable

Tension in a span of a cable: $T = \dfrac{ws}{\sin\theta}$

For the angle of 30°, the sin30 is 0.5.

Therefore the tension is: $T = \dfrac{ws}{\sin\theta} = \dfrac{2,000\ \text{N}}{0.5} = 4,000\ \text{N}$

For the equation describing the tension with an angle of 30°, we find the minimum tension T_0, where $T_0 = T\cos\theta$.

$$T_0 = T\cos\theta = 4,000\ \text{N}\ \cos30° = 4,000\ \text{N}\ (0.866) = 1.7 \times 10^3\ \text{N}$$

The equation describing the tension is: $T = T_0\sqrt{1 + \dfrac{w^2}{T_0^2}s^2}$ Adding the values from above:

The equation describing the tension is: $T = 1.7 \times 10^3\ \text{N}\sqrt{1 + \dfrac{40\ \text{N/m}}{1.7 \times 10^3\ \text{N}}s^2}$

The shape of the cable is: $y = \dfrac{T_0}{w}\left(\cosh\left(\dfrac{w}{T_0}x\right) - 1\right)$

Adding the values from above:

$$y = \dfrac{1.7 \times 10^3\ \text{N}}{40\ \text{N/m}}\left(\cosh\left(\dfrac{40\ \text{N/m}}{1.7 \times 10^3\ \text{N}}x\right) - 1\right)$$

The shape of the cable is: $y = 43.3\ \text{m}\ (\cosh(23.0 \times 10^{-3}\ /\text{m})x - 1)$

Concept Reinforcement

1. Explain the difference between a load along a straight line and a load along a cable.

2. Explain a typical hanging cable shape of a load along a cable.

3. Explain the equation for the tension in a cable under its own deadweight load.

4. State the equation for the shape of the cable.

Section 3.9 – Discrete Loads on a Cable

Section Objective

- Explain discrete loads

Introduction

This section presents the concept of the shape and tension in a cable that is subjected to various discrete loads. For the analysis the cable cannot elongate, is weightless, and cannot transmit bending moments. The analysis utilizes the standard equilibrium equations for horizontal and vertical force as well as moments even though moments cannot be transmitted. The objective is to find the horizontal and vertical locations of the point of application for the discrete loads and the forces that support the cable. A second objective is to determine the tension in the cable itself.

Discrete Loads

The forces and shape of a cable can be analyzed for several situations:

   A. when the cable is carrying a uniformly distributed load, as with a suspension bridge.

   B. when the cable itself is analyzed based on its deadweight load and no external load.

   C. when the cable is carrying several non-uniform loads.

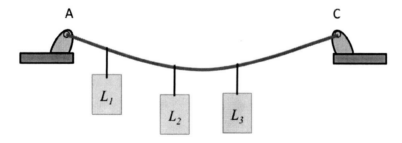

A set of discrete loads along a cable

This section analyzes the situation when the cable is carrying several non-uniform loads.

In this example, the cable is used to carry several loads. The actual deadweight load of the cable itself is typically ignored because this will be much less than the weight of the loads.

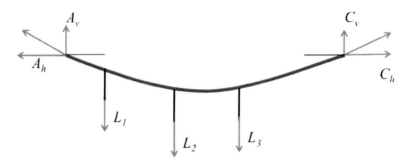

A free body diagram of a set of discrete loads along a cable

Three assumptions are used in these problems:

1. The deadweight of the cable is small enough to be ignored.

2. The cable is ultimately flexible and therefore cannot transmit a moment

3. The axial stiffness of the cable is infinite which means the cable will not elongate due to excessive loads

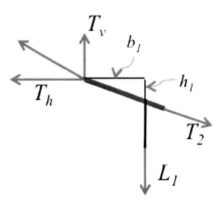

A free body diagram of a portion of the above cable

These are problems of equilibrium. Mechanical equilibrium is a condition of an object where the sum of the forces acting on the object is zero and the sum of the moments acting on the object is zero. When considering two-dimensional equilibrium we have:

Mechanical Equilibrium: $\sum F_x = 0$, $\sum F_y = 0$ and $\sum M_z = 0$.

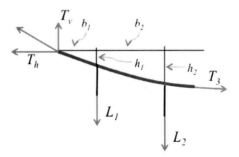

A free body diagram of a larger portion of the above cable

The vector sum of the external forces and force components in the *x*-direction is zero and the sum of the external forces and force components in the *y*-direction is zero. The sum of the moments acting on the object in the *z*-direction is zero. All moments in a two-dimensional (x, y) system act in the *z*-direction.

Even though the cable is assumed to be flexible enough that it cannot transmit moments, these equations are still valid within the context of equilibrium.

For every section of cable, the conditions of equilibrium exist and therefore a set of equilibrium equations can be constructed to solve for the variables that are unknown. Normally, the desired outcome is to know the vertical location of each load, the vertical and horizontal components of the tension at the supports, and the tension in the cable at each span (between the support and the load, between each load, etc.). As with all solutions of multiple algebraic equations, there must be one equation for each unknown. For example, if there are five unknown variables, there must be five equilibrium equations. Likewise, if there are 20 unknown variables, there must be 20 equations, and so on.

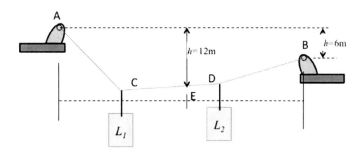

A cable with two separate discrete point loads

We will use an example problem to demonstrate this concept. The example uses a cable that has two discrete point loads.

Because some amount of information is required in order to completely solve for all unknowns, several dimension are given.

Example 1: Cable with Discrete Loads

We first construct a free-body diagram of the entire cable structure.

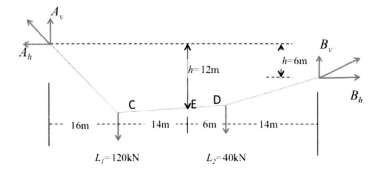

The free-body diagram of a cable with two discrete point loads

The equilibrium equations can be defined based on the free body diagram.

We examine the equilibrium at the point E, which is partway between the two loads. The system has four unknowns: A_v, A_h, B_v, B_h. Therefore, we must have four equations.

In the x-direction, $\sum F_x = 0$:

$$\sum F_x = A_h - B_h = 0$$

In the y-direction, $\Sigma F_y = 0$:

$$\sum F_y = A_v + B_v - 120 \text{ kN} - 40 \text{ kN} = 0$$

The moments in the z-direction, $\Sigma M_z = 0$:

$$\sum M_z = -A_h 6 \text{ m} - 120 \text{ kN} (34 \text{ m}) - 40 \text{ kN}(14 \text{ m}) + A_v 50 \text{ m} = 0$$

At this point, there are still four unknowns. Therefore, a fourth equation is necessary.

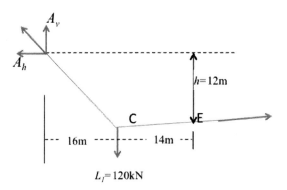

The free-body diagram of a cable with two discrete point loads
taken from a point E between the two loads

For a fourth equation, we can examine the section of the cable to the left of point E and sum the moments about the point E. This will include only two unknowns: A_v, and A_h.

$$\sum M_{zE} = -A_h 12 \text{ m} - 120 \text{ kN}(14 \text{ m}) + A_v 30 \text{ m} = 0$$

Combining the two moment equations for a solution:

$$\sum M_{zB} = -A_h 6 \text{ m} - 120 \text{ kN}(34 \text{ m}) - 40 \text{ kN}(14 \text{ m}) + A_v 50 \text{ m} = 0$$

$$-A_h 6 \text{ m} - 4.08 \times 10^3 \text{ kNm} - 560 \text{ kNm} + A_v 50 \text{ m} = 0$$

$$A_v = \frac{A_h 6 \text{ m} + 10^3 \text{ kNm}}{12 \text{ m}} = \frac{30 \text{ m}}{12 \text{ m}} A_v - \frac{1.68 \times 10^3 \text{ kNm}}{12 \text{ m}} = 2.5 A_v - 140 \text{ kN}$$

Substituting this into the moment equation about point E:

$$\sum M_{zE} = -A_h 12 \text{ m} - 120 \text{ kN}(14 \text{ m}) + A_v 30 \text{ m} = 0$$

$$-A_h 12 \text{ m} - 1.68 \times 10^3 \text{ kNm} + A_v 30 \text{ m} = 0$$

$$A_h = \frac{A_v 30 \text{ m} - 1.68 \times 10^3 \text{ kNm}}{12 \text{ m}} = \frac{30 \text{ m}}{12 \text{ m}} A_v - \frac{1.68 \times 10^3 \text{ kNm}}{12 \text{ m}} = 2.5 A_v - 140 \text{ kN}$$

The solution for A_h is:

$$A_h = 2.5 A_v - 140 \text{ kN} = 2.5 \,(0.12 A_h + 92.8 \text{ kN}) - 140 \text{ kN} = 0.3 A_h + 92 \text{ kN}$$

$$A_h (1 - 0.3) = 92 \text{ kN and } A_h = 131 \text{ kN}$$

Solving for A_v we have:

$$A_v = 0.12 A_h + 92.8 \text{ kN} = 0.12(131 \text{ kN}) + 92.8 \text{ kN} = 108.6 \text{ kN}$$

From: $\sum F_x = A_h - B_h = 0$, the solution for $B_h = 131$ kN

From: $\sum F_y = A_v + B_v - 120 \text{ kN} - 40 \text{ kN} = 0$. The solution for B_v is: 51.4 kN

This completes stage 1, finding the solution for these unknowns.

Next we determine the exact height of the two loads.

Summing the moments about point C:

$$\sum M_{zC} = -A_h(h_c) + A_v(16 \text{ m}) + 40 \text{ kN}(20 \text{ m}) - B_v(34 \text{ m}) + B_h(h_C - 6 \text{ m}) = 0$$

$$\sum M_{zC} = -131 \text{ kN}(h_C) + 108 \text{ kN}(16 \text{ m}) + 40 \text{ kN}(20 \text{ m}) - 51.4 \text{ kN}(34 \text{ m})$$
$$+ \, 131 \text{ kN}(h_C - 6 \text{ m}) = 0$$

But we can look at just the section to the left of point C, since the system is in equilibrium:

$$\sum M_{zC} = -A_h(h_c) + A_v(16 \text{ m}) = 0.$$

$$\sum M_{zC} = -131 \text{ kN}(h_C) + 108 \text{ kN}(16 \text{ m}) = 0$$

And solving for the height of point C: $h_c = \dfrac{108.6 \text{ kN}(16 \text{ m})}{131 \text{ kN}} = 13.2 \text{ m}$

The same method allows us now to solve for the height of point D. Summing the moments about point C:

$$\sum M_{zD} = -A_h(h_D) + A_v(32 \text{ m}) - 120 \text{ kN}(20 \text{ m}) = 0$$

$$\sum M_{zD} = -131 \text{ kN}(h_D) + 108.6 \text{ kN}(32 \text{ m}) - 120 \text{ kN}(20 \text{ m}) = 0$$

$$\sum M_{zD} = -131 \text{ kN}(h_D) + 3{,}475 \text{ kNm} - 2{,}400 \text{ kNm} = 0$$

$$h_D = \frac{3{,}475 \text{ kNm} - 2{,}400 \text{ kNm}}{131 \text{ kN}} = \frac{1{,}075 \text{ kN}}{131 \text{ kN}} = 8.2 \text{ m}$$

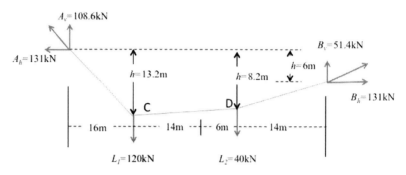

The free-body diagram of a cable with two discrete point loads

The next step is to solve for the tension in each section of the cable.

The horizontal component of the tension can be assumed to be constant at 131 kN because the load is only vertical.

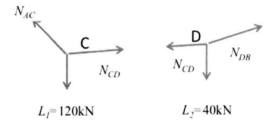

The free-body diagram of a cable with two discrete point loads
with the tension in the cable sections labeled

The formula for the horizontal section of the tension is $H = N\cos\theta$. We can rearrange this to solve for the tension (N): $N = \dfrac{H}{\cos\theta}$ where θ is the angle to the horizontal of the section of the cable. This can use the substation and give an easier formula to use:

$$N = \frac{H}{\cos\theta} = H\sqrt{1 + (\tan\theta)^2}$$

Since the tangent is a ratio of heights, we can just use the various dimension to calculate the tension.

In section AC: $\tan_{AC} = \dfrac{13.2 \text{ m}}{16 \text{ m}} = 0.825$

$$N_{AC} = H\sqrt{1+(\tan\theta)^2} = 131\,kN\sqrt{1+0.825^2} = 170\ kN$$

In section CD: $\tan_{CD} = \dfrac{5\ m}{20\ m} = 0.25$

$$N_{CD} = H\sqrt{1+(\tan\theta)^2} = 131kN\sqrt{1+0.25^2} = 135kN$$

In section DB: $\tan_{DB} = \dfrac{2.2\ m}{14\ m} = 0.157$

$$N_{DB} = H\sqrt{1+(\tan\theta)^2} = 131kN\sqrt{1+0.157^2} = 133kN$$

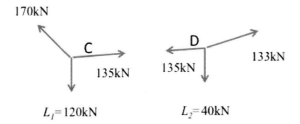

The free-body diagram of a cable with two discrete point loads
with the tension in the cable sections labeled

Concept Reinforcement

1. Explain a cable with discrete loads.

2. Explain the assumptions associated with a solving a problem of a cable with disrete loads.

3. Explain two-dimensional mechanical equilibrium.

4. Explain the equations required to solve a problem of a cable with discrete loads given the widths between points and the height of a random point on the cable.

Section 3.10 – Characteristics of Liquids and Gases

Section Objective

- Describe liquids and gases

Introduction

This section presents the concept of using liquids and gases to create loads in mechanical systems. Liquids are sufficiently massive that gravity affects the molecules which results in the liquid molecules moving downward and spreading out in a container so that the area of the container has a uniformly distributed load. Liquids do not compress and this allows the use of liquids for transmitting compressive loads from place to place in a closed system using techniques connected with hydraulics. Gases are rarely massive enough to be affected by gravity. Gases are also quite compressible. These characteristics of gases allow them to be used in closed mechanical systems as well to transmit compressive loads but not as effectively as liquids. This is the mechanical study of pneumatics.

Liquid Loads

The concept of a distributed load becomes more detailed and nuanced with the application of liquid loads or gas loads. When a liquid is placed on a horizontal surface, the liquid spreads out evenly over the surface area. Without walls to block the flow, the liquid will eventually spread everywhere. The liquid will become a thinner and thinner horizontal layer until it starts to break apart. This occurs as the forces of cohesion within the liquid, which keep the molecules together, are overtaken by the forces of separation, which are pulling the molecules apart. The liquid forms beads that roll off the surface under the influence of gravity.

If a liquid is placed in a container, the liquid spreads to take the shape of the entire container. This assumes the container is not moving and has no holes in the walls. Under the influence of gravity, the liquid molecules will fill practically any shaped container. The liquid will flow in every direction, resulting in the top surface of the volume being flat. This will happen with any container of any shape, even if the bottom surface of the container is irregular.

If the height of a container is uniform and the bottom surface is horizontal and flat, a liquid placed in this container will become a uniformly distributed load over the entire area covered by the container. This is true with any substance that exhibits the standard characteristics of a liquid, primarily being incompressible and exhibiting translational flow among the molecules.

Generally, most substances are liquid under specific conditions of temperature and pressure. Examples of common substances that are liquid at room temperature (25°C) include water, various oils, alcohol, and many chemical solutions.

To determine the total load of a defined mass of liquid, it is sufficient to know the calculation for the force of a mass that experiences gravity; i.e. gravitational acceleration.

Force due to Gravity: $F = mg$.

Where:
F = force in newtons (N)
m = mass in kilograms (kg)
g = gravitational acceleration of 9.8 meters/second squared (9.8 m/s^2)

When the mass of the liquid is spread over a defined area, the total mass will continue to exhibit the same load (force due to gravity) within the area. It is now a uniformly distributed load over an area that can be described as a defined force over a defined surface area. Force per unit surface area is commonly called pressure.

Pressure = force/area: $P = \dfrac{F}{A}$.

Where:

P = pressure in pascals (Pa) or newtons/square meter (N/m^2)
F = force in newtons (N)
A = area in square meters (m^2)

The pressure is measured as applied normal to the surface.

Liquid Properties

In liquids, the individual particles are confined by the volume of their container. Liquids flow to assume the shape of the container holding them. The shape is limited only by the volume of the liquid. The direction of the flow of liquid is primarily determined by gravitational forces. This causes a liquid to maintain a level top surface as it flows into the container. Remember that liquids do not easily compress.

Physical Properties of Liquids

The physical properties of liquids vary a lot. The properties depend upon the nature and strength of the intermolecular forces of attraction and repulsion among the particles (atoms, molecules, ions) making up the liquid.

Honey has a high viscosity at room temperature.

The viscosity of a liquid is a measure of the resistance to flow. Honey has a high viscosity at room temperature and a low viscosity at higher temperature. For a liquid to flow, the molecules must be able to slide past one another.

In general, the viscosity of a liquid increases with:

- Stronger intermolecular (between molecules) forces of attraction.

- A greater ability to form intermolecular (between molecules) bonds, especially involving several bonding sites per molecule.

- Increasing size and surface area of molecules which also increases intermolecular (between molecules) forces.

- Longer molecules, because they are more likely to get tangled up with one another which makes it harder for them to flow.

As temperature increases and the molecules move more rapidly, their kinetic energies are better able to overcome intermolecular attraction. Viscosity therefore decreases with increasing temperature.

A "thin" fluid such as water has a low viscosity and a "thick" fluid such as oil has a higher viscosity.

Material	Temp (°C)	Viscosity (μ)
Air	15	0.0000179
Water	40	0.00065
Water	20	0.001
Ethyl Alcohol	20	0.0011
Milk	25	0.003
Canola Oil	25	0.057
Olive Oil	20	0.084
Light Machine Oil	20	0.102
Molasses	20	5.
Honey	20	10.
Peanut Butter	20	150.

Viscosity can be measured in both liquids and gases and can be more technically explained as fluid friction or a measure of the resistance of a fluid to shear stress.

The various layers of the liquid experience friction as the molecules in the layers move past one another. In liquids, the viscosity is due to the cohesive forces between the molecules while in gases the viscosity is due to collisions between the molecules.

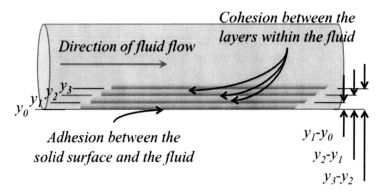

Description of friction within a fluid flow

A fluid flowing in a tube will adhere to the walls of the tube. This initial layer of the fluid, which is adhered to the wall, will travel slower than the remainder of the fluid. Because this layer travels slower it will impede the next layer of fluid through the cohesive forces of attraction between the molecules in the different layers of the fluid. Each layer impedes the next layer into the fluid until we get to the layer of molecules at the center of the tube.

The viscosity of the fluid will impact the speed with which it flows and hence the speed with which a defined liquid volume over an area will exhibit uniform distribution of the load. This also impacts how a fluid flows onto or off of a surface. A high viscosity fluid will cause greater resistance to movement and hence a force of drag or friction that is higher than a low viscosity fluid.

Cohesion and Adhesion

All liquids exhibit both cohesive and adhesive forces. Cohesive forces are forces of attraction between the molecules. Adhesive forces are forces of attraction between the molecules of the liquid and external surfaces.

Surface Tension

Molecules below the surface of a liquid are influenced by intermolecular attraction in all directions. Once at the surface, molecules orient themselves to maximize their intermolecular forces towards the fluid. This results in the surface molecules directing their cohesion forces only at the molecules below the surface within the fluid.

Another way to understand this tendency is to consider water droplets.

The forces that create this surface tension also create a preference for minimum surface area whenever possible.

This is evident when a drop of a liquid is in free fall or in free space. It will always assume a spherical shape to minimize its surface area.

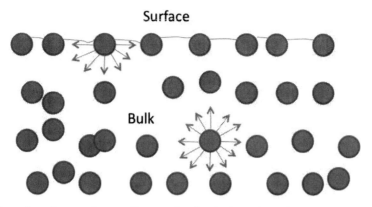

Surface

Bulk

Fluid molecules at the surface cohesively bond preferentially inward, into the bulk of the fluid.

The surface tension on water is often able to resist the pressure of small insects and light objects but can also hold up heavier metal objects, such as needles, if they are dry and carefully placed on the surface.

The surface tension has an impact when objects are moved into or out of a body of a liquid. The resistance to movement at the surface will be slightly stronger than that within the bulk of the liquid.

Capillary Action

The forces of attraction between the molecules within liquids can be characterized as cohesive forces. There are also forces of attraction to foreign materials and these are characterized as adhesive forces. Water in a glass container often "adheres" to the sides and forms a meniscus up the inside surface of the glass.

In the left tube the fluid adheres to the surface causing the molecules at the glass-water interface to rise above the surface; i.e. the capillary effect. In the right tube, there is no effect.

Adhesive forces are due to the atomic charge interaction between the molecules of the liquid and the foreign surface.

Some liquids will exhibit adhesive forces with certain materials that are stronger than their internal cohesive forces. Water does this.

Mercury is a liquid with internal cohesive forces that are much stronger than the adhesive forces between the mercury and glass. In a glass tube, mercury creates a 'reverse' meniscus where the molecules pull the liquid away from the glass.

Capillary action is evident when a narrow glass tube is lowered into a liquid that has strong adhesive forces, which pull the liquid to the inner walls of the glass tube up to a point where the weight of the liquid resists any further upward movement.

Glass coatings of silicone polymers are used to greatly reduce the adhesive forces between water and glass. Silicone polymers are in the products used on the windshields of cars to greatly improve visibility during heavy rain.

This capillary effect can produce resistive forces of drag or friction. This depends upon the charge interaction of the liquid to the surface.

Evaporation

Evaporation is the process by which molecules at the surface of a liquid break away and enter the gas phase. The kinetic energy of the molecules in liquid depends on temperature, just like in gases. A minimum level of kinetic energy must be attained for a molecule to escape the liquid phase. Therefore, the rate of evaporation increases as the temperature increases. This is because the additional heat energy that causes an increase in temperature also causes an increase in the average kinetic energy of the molecules.

Vapor Pressure

Vapor pressure is the partial pressure of vapor (gas) molecules above the surface of a liquid that is at equilibrium at a specific temperature in a confined container. Vapor pressures always increase with increasing temperature because the rate of evaporation also increases with increasing temperature.

This occurs in closed containers because vapor molecules escape the liquid phase, but cannot escape the container. As more molecules escape the liquid, these additional gas molecules exert more pressure because of the additional collisions with the inner walls of the container. These collisions result in a transfer of kinetic energy to the container wall, which causes condensation (gas molecules cooling and changing to liquid molecules) as those gas molecules lose kinetic energy and return to the liquid phase.

Easily vaporized liquids are called volatile liquids. High volatility results in a high vapor pressure. Water is the least volatile liquid and diethyl ether is the most volatile.

Therefore, if the temperature of the mechanical system is sufficient to cause extensive evaporation, evaporation of the liquid will need to be considered in the analysis of a uniformly distributed liquid load.

To be considered an ideal gas, the space the individual gas molecules (molecular space) occupy must be insignificant when compared with the space in between the gas molecules (intermolecular space). In other words, the gas molecules can have no significant volume.

The intermolecular forces between the gas molecules of an ideal gas are negligible and are disregarded in calculations. Gas molecules are usually moving so rapidly that their interactions are not affected by the intermolecular forces. Ideal gases do not exist in the natural world, but in most cases, real gases act very much like ideal gases.

Ideal Gas Law

If both of these conditions are valid, then the gas acts as an ideal gas and the Ideal Gas Equation can be used:

Ideal Gas Equation: $PV = nRT$

Where:

P = pressure in atmospheres (atm) or Pascal (Pa)
V = volume in liters (L) or cubic meters (m^3)
n = number of moles of the gas in moles (mol)
R = the Ideal Gas Constant with two values, one for (atm) and one for (Pa)
R = 0.0821 L × atm/mol × K or R = 8.3145 J/mol × K
T = temperature in Kelvin (K)

The Ideal Gas Law is a combination of various relationships that were developed between pressure, volume, and temperature acting on a gas. The relationships that were developed are:

Boyle's Law showed that volume is inversely proportional to the pressure: This is at constant temperature (T) and number of moles of the gas (n).

$$V \propto \frac{1}{P} \text{ or } PV = \text{constant or } P_1V_1 = P_2V_2$$

Charles' Law shows that volume is proportional to temperature. This is true at constant pressure (P) and number of moles of the gas (n).

$$V \propto T \text{ or } \frac{V}{T} = \text{constant or } \frac{V_1}{T_1} = \frac{V_2}{T_2}$$

Gay Lussac's Law shows that pressure is proportional to the temperature. This is at constant volume (V) and number of moles of the gas (n).

$$P \propto T \text{ or } \frac{P}{T} = \text{constant or } \frac{P_1}{T_1} = \frac{P_2}{T_2}$$

Gay Lussac also determined the **law of combining volumes** which states that at a given pressure and temperature, the volumes of gases that react with one another are in the ratios of small whole numbers.

Avogadro developed a hypothesis related to the number of molecules in a gas.

Avogadro's hypothesis: Equal volumes of gases at the same temperature and pressure contain equal numbers of molecules.

From this, Avogadro found that approximately 22.4 L (liters) of any gas at 0°C and 1 atm contained 6.02×10^{23} gas molecules, which is one mole of molecules.

Avogadro's Law followed his hypothesis and showed that volume is proportional to the number of moles at constant temperature (T) and pressure (P):

$$V \propto n \text{ or } V = \text{constant} \times n$$

Real Gases

A real gas could violate these two above conditions of the gas molecules occupying minimal space and the gas molecules exhibiting minimal intermolecular force.

This usually happens when pressures are extremely high or the temperatures are extremely low. Normal pressure is considered 1 atmosphere, the pressure on the Earth's surface on a normal day. Normal temperatures are the standard ambient temperature of about 25°C (298 K).

At extremely high pressures, the first condition is no longer valid. The space the individual gas molecules (molecular space) occupy is not insignificant when compared with the space in between the gas molecules (intermolecular space). Therefore, at these very high pressures, it is not valid to use the ideal gas equations that assume the molecules occupy no space.

At extremely low temperatures, gases are slowed down to the point that their speed does not shield them from the influence of the intermolecular forces. Again, the ideal gas equations, at these extremely low temperatures, are not valid.

Helium contained in balloons

Because of the complication of the equations needed to evaluate a real gas, chemistry courses usually work with the assumption that all gases act as ideal gases.

Concept Reinforcement

1. Explain how pressure depends on depth in a static fluid.

2. Find the static fluid pressure with $\rho=1{,}000.0$ kg/m³, $h = 0.2$ m.

3. Find the static fluid pressure with $\rho=1{,}000.0$ kg/m³, $h = 0.6$ m.

4. Explain Pascal's Principle.

5. Define the difference between an ideal gas and a real gas.

Section 3.11 – Pressure, Center of Pressure

Section Objective

- Explain pressure and center of pressure

Introduction

This section presents the concepts of pressure, static fluid pressure and the center of pressure. Pressure is the result of a load applied over a specific area and pressure is defined as a force per unit area. The static fluid pressure is the pressure that an object experienced based upon the density of a fluid and the depth within the fluid. A denser fluid has more mass per unit volume and therefore is heavier. As an object sinks deeper under the surface of the fluid the amount of fluid above it increases and hence the weight of that amount of fluid is larger. Therefore the pressure is larger. For analysis purposes the force that is spread over an area, otherwise known as pressure, can be resolved into a point load. The center of pressure is the location at which that point load is applied.

Loads and Pressure

There are many situations where a load is distributed over a wide area. The liquid sealants spread over the surface of the roof of a building or between the floors to reduce water leakage are examples. The rain that collects on that sealed roof enclosure is another. These loads are the result of gravity acting on the mass of the material, the sealant or the water.

When the wind acts on the sails of a schooner this is also a distributed load, albeit a force caused by the aerodynamics and wind.

In all these cases, we are talking about distributed loads of liquids and gases, which are generally called fluids. The surfaces are being impacted by a large volume of fluid and, in some cases, are submerged within the fluid.

The force experienced by the surface of an object due to fluids is normally called pressure, which is the force per unit area.

Pressure = force/area: $P = \dfrac{F}{A}$

Where:

P = pressure in pascals (Pa) or newtons/square meter (N/m²)
F = force in newtons (N)
A = area in square meters (m²)

The pressure is measured as applied normal to the surface.

The U.S. Customary units for pressure are "pounds per square inch" or psi.

Example 1: Pressure

Given a liquid volume that translates to a force of 100 kN, we need to determine the pressure on a roof area of 10 m² and on a roof area of 50 m².

This is an application of the formula for pressure = force/area: $P = \dfrac{F}{A}$.

For the roof area of 10 m², the pressure is: $P = \dfrac{F}{A} = \dfrac{100 \text{ kN}}{10 \text{ m}^2} = 10 \text{ kN/m}^2$.

For the roof area of 50 m², the pressure is: $P = \dfrac{F}{A} = \dfrac{100 \text{ kN}}{50 \text{ m}^2} = 2 \text{ kN/m}^2$.

If the specification for the maximum pressure on the roof is 5 kN/m², only the second situation would be approved because it is below the specification.

Atmospheric and Gauge Pressure

Standard atmospheric pressure on the surface of the Earth is 14.7 pounds per square inch (14.7 psi) in the English System and 101,325 Pascal (101,325 Pa) in the SI System. This is called one standard atmosphere.

When the pressure in a pneumatic system, for example in tires, is measured using a gauge, the measurement is the pressure above standard atmospheric pressure. Therefore, actual total pressure is the gauge pressure plus the atmospheric pressure.

Therefore, the total pressure experienced by a system that is showing a gauge pressure of 35 psi at the surface of the Earth would be 35 psi + 14.7 psi = 49.7 psi. In the SI system of units, the gauge pressure of 35 psi would be 241,250 Pa. Therefore, the total pressure would be 241,250 + 101,325 Pa = 342,575 Pa.

Center of Pressure

When the pressure is applied over a surface, the pressure is the same on each small square surface area of the object.

This is true in all three dimensions. The force per unit area acting on an object submerged in a fluid is the same at all points around the object at the same depth in the fluid.

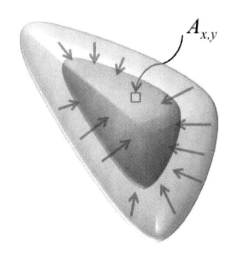

This is presented in the formula for static fluid pressure on an object placed into a fluid that is stable and which has a uniform density. At any specific height in the fluid, the pressure that is exerted on an object can be obtained at any location around the object using the following equation.

The static fluid pressure is calculated as: $P = \dfrac{\text{weight}}{\text{area}} = \dfrac{mg}{A} = \dfrac{\rho(Ah)g}{A} = \rho g h$

Where:
P = pressure in pascals (Pa), where 1 Pa = 1 N/m²
ρ (Greek letter rho) = fluid density in mass/volume (m/V) (in kg/m³)
g = gravitational acceleration g=9.8 m/s²)
h = depth of fluid in meters (m)
A = horizontal Area of the water in square meters (m²)
V = Volume = Ah in cubic meters (m³)
m = mass = ρV = density × volume in kilograms (kg)

When examining the surface of an object and considering the load applied by a fluid on that surface, we can resolve the distributed load into a single load at a point similar to resolving a set of loads acting on a beam or a set of loads acting on a surface.

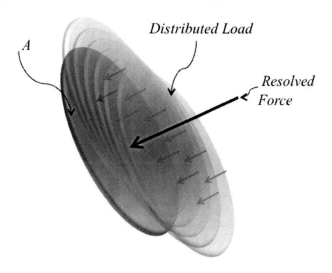

The point at which a concentrated load applied to the surface is comparable to the distributed load of the fluid is called the Center of Pressure.

For example, if the fluid is in an evenly distributed layer over a horizontal surface that is a perfect circle, the center of pressure will also be the center of the circle.

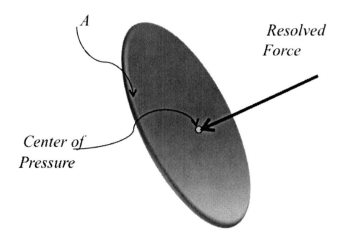

Example 2: Center of Pressure

A uniform layer of fluid is covering a vast area that is the exact shape of a right triangle with the base length of 10 m and the height of 20 m. Determine the center of pressure.

This uses the same concepts as the centroid of the area of a triangle, which can be obtained from a book of formulas. The centroid of the area of a right triangle is located at 1/3 the base (in the x-direction) and 1/3 the height (in the y-direction).

Therefore, the centroid of the area is located at 10/3 m from the edge of the base (the edge closer to the right angle) and 10/3 m of the height from the bottom. This is also the center of pressure.

Concept Reinforcement

1. Explain two examples of typical fluid pressure applications.

2. State the standard units for pressure in SI Units and Customary Units.

3. Write the formula for pressure.

4. Given a force of 40 N and a surface area of 10 m², define the pressure acting on the surface.

5. Given a load of 150 kN of fluid acting on a perfect circle of surface area 20 m², define the center of pressure and the resolved load acting at that point.

Section 3.12 – Pressure in a Stationary Liquid

Section Objective

- Describe pressure in a stationary liquid

Introduction

This section presents the concept of pressure in a stationary fluid. Most liquid closed systems that use pressure for transmitting force are considered to have stationary liquid. Within a stationary liquid, the pressure is generally dependent upon the depth within the fluid. For any objects that are submerged into a liquid the pressure that the object experiences is dependent upon the depth within the fluid. Another aspect to fluids and pressures is that external forces can be multiplied through a hydraulic system just by changing the area exposed to the pressure. Any external pressure is experienced uniformly throughout the liquid and therefore a force applied to a small area translates to a pressure. This pressure, when exposed to a larger area translates to a larger force.

Pressure in a stationary liquid

Pressure, which is a force exerted over a surface (force per unit area), can be transmitted through a fluid and is felt equally throughout the entire fluid. Because of the mobility of the molecules in a fluid, fluids are able to exert pressure in any shape container and are able to transmit pressure over long distances.

In a static (non-moving) fluid, the pressure exerted depends on how far below the surface of the fluid the pressure is measured, the density of the fluid and the gravitational acceleration. The pressure in a static fluid results from the weight of the fluid and does not depend upon the shape of the container, the total mass of the fluid, or surface area of the fluid.

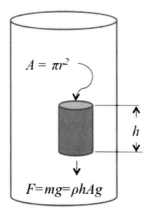

To determine the weight of the fluid, you must first determine the mass. Density is a measure of mass per unit volume in kilograms per cubic meter (kg/m³). The volume of the fluid is usually more easily established using the dimensions of the container.

Consider a fictional container that is massless, shaped like a column and that exists within a volume of a fluid (see the figure). This container holds a fraction of the fluid, a volume that is defined by an area of A and the height of h. The volume of the fluid contained in this fictional column is $V = Ah$.

Returning to the determination of the mass, we have a fluid with a known density and now we have the volume. The mass is then equal to the density times the volume, as represented by the formula $m = \rho V$.

The **weight of the fluid** that is contained within this massless container is:

$$\text{Weight} = ma = mg = (\rho V)g = (\rho Ah)g$$

The **static fluid pressure** is calculated as $P = \dfrac{\text{weight}}{\text{area}} = \dfrac{mg}{A} = \dfrac{\rho(Ah)g}{A} = \rho gh$

Where:
P = pressure in pascals (Pa), where 1 Pa = 1 N/m²
ρ (Greek letter rho) = fluid density in mass/volume(m/V) (in kg/m³)
g = gravitational acceleration $g = 9.8$ m/s²)
h = depth of fluid in meters (m)
A = horizontal Area of the water in square meters (m²)
V = Volume = Ah in cubic meters (m³)
m = mass = ρV = density × volume in kilograms (kg)

Note that for this equation to provide accurate answers the fluid is assumed to be incompressible. This is an accurate assumption for most fluids, including water.

Notice that in the equation the static fluid pressure depends only on the density of the fluid, the gravitational acceleration and the height of the fluid above the measuring point. Though it does depend upon the mass and the volume of the fluid these are both presented in the formula through the use of density. Therefore the static fluid pressure is the same for a specific depth within any volume or mass of the fluid. This means that no conceivable geometric shape for the container changes the basic fact that at a specific depth the pressure is the same everywhere. It is independent of the shape.

Example 1: Static Fluid Pressure

A container with an area (A) of 60 cm² has water filled to a height of 30 cm. the density of water is 1.00 gr/cm³. Determine the static fluid pressure at the top and at the bottom of the fluid.

At the top of the container, the static fluid pressure uses the same formula $P = \rho gh$. For a height of zero, the static fluid pressure at the surface of the water is also zero.

At the bottom of the fluid, the height is 30cm. We use the same formula: $P = \rho gh$.

However, we change the value for density to kilograms per cubic meter.

1 gram = 0.001 kg = 1×10⁻³ kg. 1 cm = 0.01m and therefore if we cube each side we get 1 cm³ = 0.0000001 m³ = 1×10⁻⁶ m³.

density of water = 1 gr/cm³ = 0.001 kg/0.000001 m³

density of water = 1 × 10⁻³ kg/1 × 10⁻⁶ m³ = 1,000 kg/m³

The height of the water is 30 cm = 0.3m. Returning to our equation:

$$P = \rho g h = (1,000 \text{ kg/m}^3)\, 9.8 \text{ m/s}^2\, (0.30 \text{ m}) = 2,940 \text{ kg/m} \times \text{s}^2$$

To get to the units for pressure of newtons per square meter, we need to make a slight modification. We multiply the units of the numerator and denominator by m.

$$P = 1,960 \text{ kg/m} \times \text{s}^2 = \left(2,940 \frac{\text{kg}}{\text{m} \times \text{s}^2}\right)\left(\frac{\text{m}}{\text{m}}\right) = \frac{2,940 \text{ kg/m} \times \text{s}^2}{\text{m}^2} = 2,940 \text{ N/m}^2$$

Pascal's Principle

Pascal's Principle states

"A change in the pressure of an enclosed incompressible fluid is conveyed undiminished to every part of the fluid and to the surfaces of its container."

Pressure is force/area and is presented in newtons per square meter (N/m²). The Pascal principle shows that if the pressure is transmitted through the fluid, the force that is applied only depends on the surface area of the fluid to which it is applied. The additional pressure is not dependent upon the volume, density or depth. The increase in pressure is uniform throughout the volume of the fluid.

If a force of 20 N is applied over an area of 0.02 m² of a fluid, the pressure is 1,000 N/m². When we examine another section of a surface of 4 m², the force applied is 4,000 N. The pressure is the same because 4,000 N/4m² = 1,000 N/m².

Therefore, with a small force applied over a small area (also called pressure) and an incompressible fluid, we can obtain a substantially larger force that can be utilized to do work. This is how hydraulic pressure systems work.

This calculation assumes the height of the applied pressure is the same. An example is a U-shaped system, where the initial force is applied at the top of one arm and the resultant force is available at the top of the other arm.

In the case where the resultant force is available at a different depth, we would need to add the static fluid pressure $P = \rho g h$ to the calculations.

Example 1: Pascal's Principle

In the U-shaped configuration in the figure, a pressure of 200 kPa is applied downward to the narrow tube at the left, which has 15 cm² of surface. The right tube is 3 times the area of the left tube. Calculate the pressure and force at the surface of the right tube.

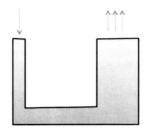

Using Pascal's principle, we know the pressure experienced at the surface of the fluid in the large right tube is the same as that experienced in the narrow left tube. Therefore, the pressure is 200 kPa.

The force that must be applied to the left tube in order to obtain a 200 kPa pressure over a 10 cm² area is calculated as:

$$A_i = 15 \text{ cm}^2 = 15\,(1 \times 10^{-4} \text{ m}^2) = 1.5 \times 10^{-3} \text{m}^2$$

$P = \dfrac{F}{A}$ and $F = PA$, at the input: $F_i = PA_i = (200 \text{ kPa})(1.5 \times 10^{-3} \text{ m}^2) = 300 \text{ N}$

At the output, we have 3 times the area, therefore:

$$A_o = 45 \text{ cm}^2 = 45\,(1 \times 10^{-4} \text{ m}^2) = 4.5 \times 10^{-3} \text{m}^2$$

This means that the force experienced at the output is:

$$F_o = PA_o = (200 \text{ kPa})(4.5 \times 10^{-3} \text{ m}^2) = 900 \text{ N}$$

Concept Reinforcement

1. Explain how pressure depends on depth in a static fluid.

2. Find the static fluid pressure with ρ=1,000.0 kg/m³, $h = 0.6$ m.

3. Explain Pascal's Principle.

4. Describe the convention for angles in mechanical systems.

Section 3.13 – Virtual Work

Section Objective

- Explain virtual work

Introduction

This section presents the concept of virtual work and virtual displacement. Virtual work is a tool used in equilibrium analysis to determine equilibrium conditions based on the potential for displacement, which is the virtual displacement. In a system such as the slider-crank mechanism a specific set of forces are defined and the objective is to find the position of the various members of the system to have the system in equilibrium.

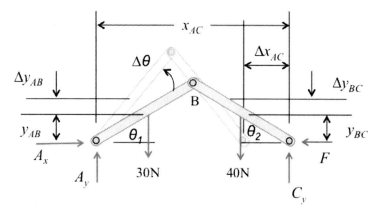

Slider-crank mechanism

Real Displacement

In our real world the idea of a displacement is a change of position between time zero (t_0) and some later time (t_1). This change in position is a necessary condition for real work to be performed.

Real Work Formulation

The basic engineering explanation of work is a force applied to an object that causes that object to move in the direction of the force. If there is no movement in the same direction as the force there is no work. For example, with circular motion the centripetal acceleration acting on an object produces centripetal force perpendicular to the direction of movement and hence there is no work performed.

Work: $W = Fd$

Where:
W = work in joules (J) or newton-meter (Nm)
F = force in newtons (N)
d = displacement in meters (m)

With real work, there is a real force and a real displacement. The force can be measured and the displacement can be measured.

Virtual Displacement

In virtual displacement, we establish a random coordinate point in space. We imagine that the point has infinitesimal changes in position which are arbitrary in direction and unpredictable in time. This is virtual displacement and can literally be just imagined. It is a theoretical tool of analysis for engineering studies.

This conceptual problem can be further expanded by defining some constraints on the movement of our point in space. For example, we can place two points in space and define that there is a distance between the two points that cannot change. This would be as if there were a tether connecting the two points so as one moved the other would also move in the same direction or in an angular movement such that the distance between the points stays the same.

Therefore the virtual displacement at this stage would be the infinitesimal changes in position of the point given the constraint that the distance to the other point cannot change.

Virtual Work

Virtual work is the work performed based upon the "potential" virtual displacement. This is used in mechanical statics to solve equilibrium problems.

Virtual Work: $W_v = Fd_v$

Where:
W_v = virtual work in joules (J) or newton-meter (Nm)
F = force in newtons (N)
d_v = virtual displacement in meters (m)

Equilibrium Conditions

On a particle:

The necessary and sufficient condition for the equilibrium of a particle is that zero virtual work is done by all the working forces acting on the object during any virtual displacement. This must be consistent with any constraints imposed on the particle – as discussed above.

On a rigid body:

The necessary and sufficient condition for the equilibrium of a rigid body is that zero virtual work is done by all the external forces acting on the object during any virtual displacement. This must be consistent with any constraints imposed on the particle – as discussed above.

No virtual work is done by internal forces, by reactions in smooth constraints or by any forces perpendicular to the direction of motion. Virtual work is done only in reactions when friction exists.

Applications to Dynamic Equilibrium

The primary application of virtual displacements is in dynamic equilibrium problems where there is an "anticipated" displacement. Therefore the equations of equilibrium are applied using infinitesimal displacement changes. A classic demonstration vehicle for virtual work is the typical crank and piston system that is common in combustion engines and pumps. This is called the "slider-crank mechanism" which converts rotary motion into back and forth motion in a piston housing.

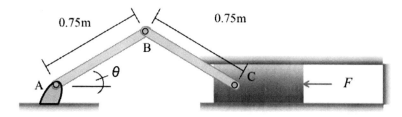

Slider-crank mechanism

Example 1: Virtual Work

A slider-crank mechanism like the one in the figure is being analyzed. The shafts are pinned at points A, B and C. The length of the shaft AB is the same as the length of the shaft BC, 0.75 m. The weight of shaft AB is 30 N and the weight of the shaft BC is 40 N. This weight applies a force to move the piston to the right. The force (F) that is opposing this is 80 N. The objective is to determine the angle θ for the slider-crank mechanism to be in equilibrium.

Solution

The first step is to create a free-body diagram. This is a free-body diagram of a dynamic system and therefore it presents the changes (Δ) in the movement in both the x and y directions.

A short explanation of the free-body diagram is necessary.

The origin is defined as point A. Therefore all the changes in position are relative to point A.

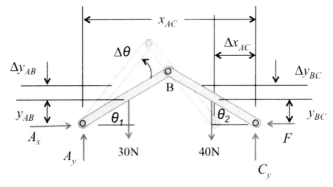

Slider-crank mechanism

Angles

The angle θ that each of the shafts experiences with respect to horizontal will always be the same. $(\theta_1 = \theta_2)$

Horizontal dimensions
x_{AC} = the original length
Δx_{AC} = the change in length (virtual displacement in the $-x$ direction)

Vertical dimensions
y_{AB} = the original height of the centroid for the shaft AB
Δy_{AB} = the change in the height of the centroid for the shaft AB (virtual displacement)
y_{BC} = the original height of the centroid for the shaft BC
Δy_{BC} = the change in the height of the centroid for the shaft BC (virtual displacement)

Forces

A_x = the virtual force exerted by the pin support on the shaft AB in the x-direction
A_y = the virtual force exerted by the pin support on the shaft AB in the y-direction

NOTE: The two forces at A perform no work because there is no displacement

F = the force exerted opposing the movement of the piston
C_y = the virtual force exerted by the pin and piston support on the shaft BC in the y-direction

NOTE: The vertical force at C performs no work because there is no vertical displacement

The loads of the two shafts (30 N, 40 N) are from the centroid of the respective shaft

It is necessary to express all the positions in terms of the angle θ. To arrive at the "Δ" terms for the different dimensions, the derivations have been omitted.

$x_{AC} = 0.5\cos\theta + 0.5\cos\theta = \cos\theta$

$\Delta x_{AC} = (\sin\theta)\Delta\theta$

$y_{AB} = 0.25\sin\theta$

$\Delta y_{AB} = 0.25(\cos\theta)\Delta\theta$

$y_{BC} = 0.25\sin\theta$

$\Delta y_{BC} = 0.25(\cos\theta)\Delta\theta$

From the free-body diagram, an increase in the angle θ of $\Delta\theta$ causes a decrease in the length of x_{AC} and an increase in the heights of y_{AB} and y_{BC}.

The virtual work done through this virtual displacement is the sum of the virtual work done by the force (F) and the virtual work done by the weight of the shafts.

$$\Delta W = 30 \text{ N}(\Delta y_{AB}) - 40 \text{ N}(\Delta y_{BC}) - F(\Delta x_{AC}) = 0$$

For equilibrium to exist, the sum of the virtual work must be zero.

We replace the terms with the formulas that include the angle θ.

$$\Delta W = 30 \text{ N}(0.25(\cos\theta)\Delta\theta) - 40 \text{ N}(0.25(\cos\theta)\Delta\theta) - 80 \text{ N}(\sin\theta)\Delta\theta = 0$$

All the signs can be changed because they sum to zero:

$$\Delta W = +30 \text{ N}(0.25(\cos\theta)\Delta\theta) + 40 \text{ N}(0.25(\cos\theta)\Delta\theta) + 80 \text{ N}(\sin\theta)\Delta\theta = 0$$

It is necessary to note that $\Delta\theta$ cannot be zero.

This simplifies to: $(7.5\cos\theta + 10\cos\theta)\Delta\theta - 80 \text{ N}(\sin\theta)\Delta\theta = 0$

This gives: $(7.5 \text{ N} + 10 \text{ N})\cos\theta\Delta\theta = 80 \text{ N} \sin\theta\Delta\theta$

This can be simplified to $(17.5 \text{ N}) \cos\theta = 80 \text{ N} \sin\theta$

and further simplified to: $\dfrac{\sin\theta}{\cos\theta} = \dfrac{17.5 \text{ N}}{80 \text{ N}}$ and the $\tan\theta = \dfrac{\sin\theta}{\cos\theta} = \dfrac{17.5 \text{ N}}{80 \text{ N}} = 0.218$.

Solving for θ we have: $\theta = \tan^{-1}(0.218) = 12.34°$.

Therefore, the equilibrium condition requires a positive angle θ of 12.5°.

Concept Reinforcement

1. Explain the concept of real displacement.

2. Explain real work and the formula for work.

3. Calculate the work done by a force of 30 N moving an object 50 m.

4. Explain the concept of virtual displacement.

5. Describe the necessary and sufficient conditions for virtual work.

Section 3.14 – Potential Energy

Section Objective

- Describe potential energy and its importance

Introduction

This section presents the concept of potential energy, which is the energy of position. In a gravitational field objects increase their potential energy by moving in the direction opposite the direction that the field would naturally pull the object. On Earth, this means raising the object up. There is also elastic potential energy which is the energy involved in spring motion where an object is repeatedly moved from one position by the spring to another position and back again. These endpoints are where the bob has the maximum potential energy. There is also a universal gravitational potential energy formula for stellar objects in space.

Potential Energy

Potential energy is a called the energy of position. This is energy that an object has because of its physical position in a field such as a gravitational field, or by virtue of its position in the reflex action of a spring.

Gravitational Potential Energy (near Earth)

Potential energy is described in several ways:

1. as energy stored in a system.

2. as the energy an object has because of its position in a force field (ex. Gravitational force field).

3. as a restoring force.

Gravitational potential energy uses the formula $U = mgh$

where
U = potential energy in joules (J)
m = mass in kilograms (kg)
g = the gravitational acceleration of 9.8 meters/second squared (9.8 m/s²)
h = the height above the zero position in meters (m)

Note: $1 \text{ J} = 1 \text{ Nm} = 1 \text{ (kg} \times \text{m/s}^2\text{)m}$
(1 joule = 1 newton-meter = 1 (kilogram-meter/second²)(meter)

Near the surface of the Earth, gravitational acceleration (g) is a constant at 9.8 m/s² (32 ft/s²). With mass (m) also being a constant under normal conditions the scale of the potential energy depends only on height.

Therefore $U = m(9.8 \text{ m/s}^2)h$ and where mass and gravity are constant the potential energy is dependent upon only the height above a location.

In scenario (a) of the illustration, the ball was moved up to a physical height from which the ball can drop. The floor level is considered the zero position. Therefore, the height of 3 m is the height used in the calculation of potential energy. To get the ball to the height of 3 m, an amount of work must be performed on the ball by a force applied over a distance ($W=Fd$, where $d=h$), which is then equal to the potential energy at that height (h).

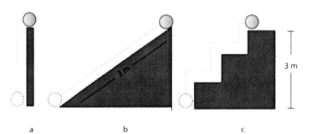

Because the calculation of potential energy in each case requires that only the vertical component of the force be considered, and not the horizontal component, the potential energy (U) is the same for each of these situations. The kinetic energy of the objects is affected by the travel paths because they move down to the ground, but all start with the same potential energy. This is because the gravitational field and the direction of gravitational acceleration are vertically down.

Example 1: Potential Energy

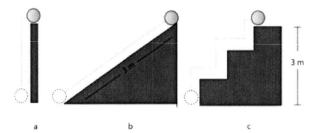

For an object with a mass of 30 kg, calculate the potential energy for each of the three scenarios:

Scenario a: height is 3 m. Gravitational acceleration is 9.8 m/s².

$U = mgh = 30 \text{ kg}(9.8 \text{ m/s}^2)3 \text{ m} = 882 \text{ J}$

Scenario b: height is 3 m. Gravitational acceleration is 9.8 m/s²

$U = mgh = 30 \text{ kg}(9.8 \text{ m/s}^2)3 \text{ m} = 882 \text{ J}$

Scenario c: height is 3 m. Gravitational acceleration is 9.8 m/s²

$U = mgh = 30 \text{ kg}(9.8 \text{ m/s}^2)3 \text{ m} = 882 \text{ J}$

There is no effect by the horizontal displacement of the path taken to arrive at the position of 3 m high. This is one result of gravitational potential energy being a conservative force.

Example 2: Potential Energy

Now, looking at scenario c and assuming the potential energy at the top of the steps is 882 J, what is the potential energy at each step?

Each step is 1 meter and there are 3 steps. At the floor the potential energy is zero. Therefore, the potential energy at each step is 1/3 of the potential energy of the top.

At the first (lowest) step, the height (h) is 1 m.

$U = mgh = 30 \text{ kg}(9.8 \text{ m/s}^2)1 \text{ m} = 294 \text{ J}$

At the second step the height (h) is 2 m.

$U = mgh = 30 \text{ kg}(9.8 \text{ m/s}^2)2 \text{ m} = 288 \text{ J}$

At the third step the height (h) is 3 m.

$U = mgh = 30 \text{ kg}(9.8 \text{ m/s}^2)3 \text{ m} = 882 \text{ J}$

In the situation where an object is held above a cliff, the potential energy is dependent upon the height of the cliff above the ground below. Initially, we stated that the ground position is assumed to be the ground of the Earth. However, the position of the "ground" level is arbitrary and can be chosen for convenience. In this case, the potential energy is determined for the difference in position and not for the absolute position about a known zero point, which is normally the Earth. When the position of the object in the gravitational field (the gravitational field is a vector; i.e. it has a direction) indicates that when the object is released, it will travel in the direction of the gravitational field, then the potential energy is positive.

Here, even if the grassy spot where the woman is standing would normally be considered at the ground level, we can set the "ground" level at the base of the cliff for convenience in calculation. The ball is able to move, once released, in the direction of the gravitational field (i.e. down) towards ground. It therefore has potential energy, which is equal to *mgh*, with *h* being the height from the base (bottom) of the cliff up to the woman's hand.

Example 3: Potential Energy

For an object of 8 kg, if the woman is standing 400 m above the base of the cliff and she holds the ball 1.5 m above where she is standing what is the potential energy?

$U = mgh = 8$ kg$(9.8$ m/s^2 $) 400$ m $= 31,360$ J?

But wait, this is the potential energy of the object at the cliff level, after it has left her hand (and dropped 1.5 m). The actual potential energy of the object while in her hand is:

$U = mgh = 8$ kg$(9.8$ m/s$^2) 401.5$ m $= 31,477.6$ J

During that 1.5 m drop from her hand to the top of the cliff, 117.6 J of potential energy are converted to kinetic energy.

Calculating Potential Energy

A more universal definition of potential energy is: the change in potential energy (associated with a conservative force) is the work done by that conservative force; i.e. $W = \Delta U$; Work = change in potential energy.

The standard formula for work is *W=Fd*.

Therefore, when doing work in a gravitational field, this can be considered performing work on the object in the vertical direction against gravity.

W=Fd = Fh = mah and for *a = g* this changes to $W = mgh$.

This formula ($W = mgh$) is the formula for potential energy in the special case of gravitational potential energy.

Common situations where potential energy occurs are the elastic potential energy, the general gravitational potential energy and the electromagnetic potential energy.

Elastic Potential Energy

Starting with potential energy equal to the work done: i.e. *W=Fd*.

The Force (*F*) on a spring: $F = kx$,

Where:
F = the force in newtons (N)
x = the distance in meters (m)
k = called the spring constant, a relative measure of the elasticity of the spring in newtons/meter (N/m).

Therefore, $W = kxd$ (but $d = x$) and the distance must be averaged, which adds a factor of ½. This gives the elastic potential energy formula:

$$U = \frac{1}{2}kx^2$$

Where:
U = elastic potential energy in joules (J)
x = the distance in meters (m)
k = the spring constant, a relative measure of the elasticity of the spring in newtons/meter (N/m).

Example 4: Elastic Potential Energy

If 60 N of force causes the spring to stretch 0.3 m, we use $F=kx$ to find the spring constant.

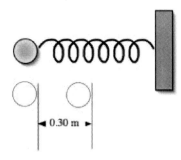

Solving for the spring constant we have:

$$k = \frac{F}{x} \text{ and } k = \frac{F}{x} = \frac{60 \text{ N}}{0.3 \text{ m}} = 200 \text{ N/m}$$

The formula for elastic potential energy is $U = \frac{1}{2}kx^2$

$$U = \frac{1}{2}kx^2 = \frac{1}{2}(200\text{N/m})(0.3\text{m})^2 = \frac{1}{2}(200\text{N/m})(0.09\text{m}^2) = 9.0\text{Nm} = 9.0\text{J}$$

General Gravitational Potential Energy

For general gravitational potential energy there is a formula that is more widely applicable than the formula U = mgh.

Gravitational Potential Energy: $U_G = -G\frac{m_1 m_2}{r}$

Where:
U = potential energy in joules (J)
G = the universal gravitational constant:

$G = 6.67 \times 10^{-11} \text{ N} \times \text{m}^2/\text{kg}^2 = 6.67 \times 10^{-11} \text{ m}^3/\text{kg} \times \text{s}^2$

m_1 and m_2 are the masses of two objects in kilograms (kg)

r = the distance between the two object in meters (m)

This applies to objects at a great distance from the earth where the force of gravity is no longer constant.

Example 5: Universal Gravitational Potential Energy

A person on the surface of the Earth has a mass (m_1) of 150 kg. The mass of the Earth (m_2) is 5.98×10^{24} kg. The distance r is to the center of the Earth, assuming the person is standing on the surface of the planet.

$$U_G = -G\frac{m_1 m_2}{r} = -6.67 \times 10^{-11}\,\text{N} \cdot \text{m}^2/\text{kg}^2 \left(\frac{(150\text{kg})(5.98 \times 10^{24}\text{kg})}{6,378,000\text{m}} \right)$$

$$U_G = -G\frac{m_1 m_2}{r} = -6.67 \times 10^{-11}\,\text{N} \cdot \text{m}^2/\text{kg}^2 \left(\frac{(8.97 \times 10^{26}\text{kg})}{6,378,000\text{m}} \right)$$

$$U_G = -9.38 \times 10^9\,\text{J}$$

Note that using the formula for potential energy $U=mgh$ would give us a value for potential energy of zero because the height (h) is zero at the surface of the Earth. Also note that the value for universal potential energy is negative.

Example 6: Universal Gravitation Potential Energy

An object is 5 million meters from the surface of the Earth. The object has a mass (m_1) of 150 kg. The mass of the Earth (m_2) is 5.98×10^{24} kg. The distance r is to the center of the Earth and therefore we need to add the radius of the Earth to the distance r.

$$r = 5,000,000 \text{ m} + 6,378,000 \text{ m} = 8,378,000 \text{ m}$$

$$U_G = -G\frac{m_1 m_2}{r} = -6.67 \times 10^{-11}\,\text{N} \cdot \text{m}^2/\text{kg}^2 \left(\frac{(150\text{kg})(5.98 \times 10^{24}\text{kg})}{11,378,000\text{m}} \right)$$

$$U_G = -G\frac{m_1 m_2}{r} = -6.67 \times 10^{-11}\,\text{N} \cdot \text{m}^2/\text{kg}^2 \left(\frac{8.97 \times 10^{26}\text{kg}}{11,378,000\text{m}} \right)$$

$$U_G = -5.26 \times 10^9\,\text{J}$$

In this calculation, at a distance 5 million meters from the surface of the Earth, the object of 150 kg experiences smaller potential energy compared to the level at the surface of the Earth. Therefore, the quantity of the potential energy decreases as we move further away from the planet.

Concept Reinforcement

1. Explain how potential energy is the energy of position.

2. Explain gravitational potential energy.

3. Explain elastic potential energy.

4. State the formula for general gravitational potential energy.

5. For an object of 40 kN determine the gravitational potential energy at a height of 20 m and at a height of 130 m.

Section 3.15 – Applications

Section Objective

- List applications of mechanical engineering

Introduction

This section presents various application; friction, a load on a cable and the static pressure in a liquid.

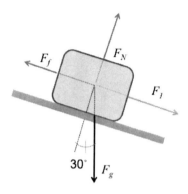

Friction Applications

The weight of a block is 1,200 N. The block is on a flat surface and is experiencing an external force of 900 N at an angle of 30° with the horizontal. The coefficient of static friction of 0.78 and a coefficient of kinetic friction of 0.45.

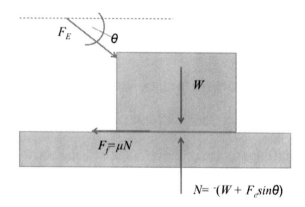

Force of static friction: $F_{fric} = \mu_s N$

The magnitude of the normal force is the sum of the weight of the block plus the fraction of the external force that is vertically down; $N = 1,200 \text{ N} + 600 \text{ N} \sin\theta$. With the angle of 30° the normal force is:

$N = 1,200 \text{ N} + 600 \text{ N} \sin 30 = 1,200 \text{ N} + 600 \text{ N} (0.5) = 1,500 \text{ N}$

The force of static friction is: $F_{fric} = \mu_s N = (0.78)(1,500 \text{ N}) = 1,170 \text{ N}$

This is the horizontal component of the required force; $F\cos\theta = 1,170 \text{ N}$

Therefore, the amount of force required to exceed the force of friction is:

$$F = \frac{-1,170 \text{ N}}{\cos 30°} = \frac{-1,170 \text{ N}}{0.866} = 1,351 \text{ N}$$

In order to get this block to move, 1,351 N of external force must be applied at an angle of 30°.

Once the block is moving, the kinetic friction is the resistive force.

$F_{fric} = \mu_k N = (0.45)(1,500) = 675 \text{ N}.$

Again, this is the horizontal component of the required force; $F\cos\theta = 675 \text{ N}$.

Therefore, the amount of force required to exceed the force of friction is:

$$F = \frac{-675 \text{ N}}{\cos 30°} = \frac{-675 \text{ N}}{0.866} = 779 \text{ N}$$

In order to keep this block moving, 779 N of external force must be applied at an angle of 30°.

Shear Force and Bending Moments

Here we examine how to quantify the shear forces and bending moments using an example. The steps that are followed to obtain the solution are the following:

Step 1: Examine the description of the problem.

Step 2: Create a free-body diagram for the problem.

Step 3: Draw the Shear Force and Bending Moment Diagram

Step 4: Define the relationships that exist in formulas with the aim to have as many equations as there are unknown elements (forces and moments).

The length from point A to point C is 40 m and from point A to point B is 16 m. The load is 50 kN.

Step 1: Description of the Problem

This description of the problem shows that the load is off center to the left, which indicates that more of the load is supported by the left support. This additional force will be proportional to the fraction of the length of the beam.

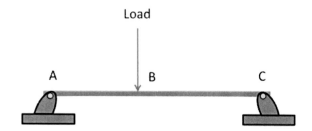

Load

A   B   C

A Horizontal beam with a vertical load at point B
experiences shear forces and bending moments

Step 2: Free-Body Diagram

To create the free-body diagram, we know that under equilibrium conditions, the vertical load is countered by an equal amount of force in the opposite direction, which is the sum of the upward forces from the two supports. This assumes that the beam itself does not deform and only transfers the forces to the supports.

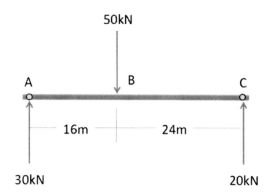

50kN

A   B   C

16m   24m

30kN   20kN

Free-body diagram of the beam

Support Force on point A: $F_A = 50 \text{ kN} \left(\dfrac{24 \text{ m}}{40 \text{ m}} \right) = 30 \text{ kN}$

Support Force on point C: $F_C = 50 \text{ kN} \left(\dfrac{16 \text{ m}}{40 \text{ m}} \right) = 20 \text{ kN}$

Step 3: Shear Forces and Bending Moments

To understand the shear forces and the bending moments, we examine a portion of the beam to the left of the load.

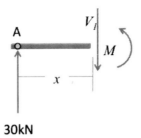

30kN

Free-body diagram of a portion of the beam to the left of the load

The entire beam is in equilibrium and therefore each section of the beam is in equilibrium. This is the basis of the determination of the shear forces and the bending moments.

Step 4: Define the equations

The beam at a distance "x" from point A experiences a shear force (V) and a bending moment (M).

To determine the forces, we again turn to the fact that the vertical forces are in equilibrium and therefore sum to zero. Therefore, the portion of the vertical load is equal and opposite to the force upward that the beam experiences from the left support.

The portion of the load is 30 kN; i.e. V_1 = 30 kN with the direction downward.

The bending moment the beam experiences is determined by the fact that the moments about the left end must also sum to zero.

Using clockwise as positive, we have:

Sum of moments: $V_1 x - M = 0$ and therefore $M = V_1 x$.

Including that V_1 = 30 kN we have: $M = (30\ kN)x$.

For example, if the distance x = 7 m the solution is $M = (30\ kN)7\ m = 210\ kNm$.

Now we examine a section of the beam on the other side of the load.

The calculation of the shear force includes the entire load as well as the support force at point A.

The equilibrium equation for the forces is:

30 kN – 50 kN – V_2 = 0; i.e. V_2 = (30-50) kN = 20 kN

Therefore, the shear force (V_2) is 20 kN upward.

The bending moment that the beam experiences is again determined by the fact that the moments about the left end at point A must also sum to zero. Using clockwise as positive, we again have the equilibrium equation for the moments:

Sum of moments: $(30\ kN)(16\ m)+V_2 x - M = 0$ and therefore $M = 480\ kNm + V_2 x$

Including that $V_2 = 20$ kN we have: $M = 480$ kNm + (20 kN)x= 480 kNm 20 kN(x)

For example, if the distance $x = 17$ m the solution is $M = 480$ kNm + (20 kN)(17 m) = 480 kNm 340 kNm = 140 kNm.

This infers that the direction shown for the moment (counterclockwise) is correct and the moment is 140 kNm in the counterclockwise direction.

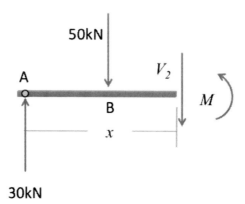

Free-body diagram of a portion of the beam to the right of the load

The Shear Force and Bending Moment Diagram is constructed from this data.

Shear Force and Bending Moment Diagram

Section 1: $0 < x < 16$ m
$V_1 = 30$ kN
$M_1 = 30$ kNx

Section 2: $16 < x < 40$ m
$V_2 = 20$ kN
$M_2 = 480$ kNm -20kNx

Shear force and bending moment equations describing the two sections

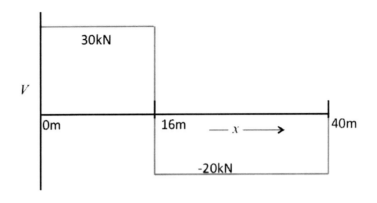

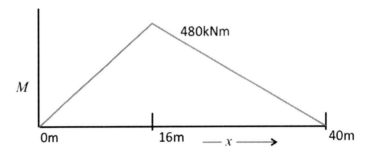

Shear Force Diagram (above) and Bending Moment Diagram (below)

The shear force (V) is constant at 30 kN from the left support to the load and is constant at 20 kN from the load to the right support.

The bending moment increases linearly from zero (0) at the left support to a maximum of 480 kNm at the location of the load (at 16 m), and decreases linearly from the location of the load to the right support.

Load along a straight line

A bridge with a uniform load (w) of 250 MN/m (250×10^6 newtons/meter) is suspended by two cables. For a section of the bridge that is 50 m long define the equations that describe the tension in the cable and the shape of the cable. Assume the angle $\theta = 30°$.

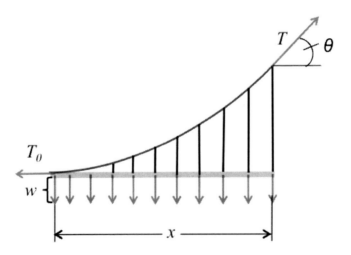

A load along a section of a suspension bridge

First, the total load is the uniform load (w) measured over the length (x).

$$wx = 250 \text{ MN/m}(50 \text{ m}) = 12.5 \times 10^9 \text{ N}$$

To determine the value of the tension T, we use the angle is 30°, where the sin30° is 0.5.

Therefore, the tension is: $T = \dfrac{wx}{\sin\theta} = \dfrac{12.5 \times 10^9 \text{ N}}{0.5} = 25.0 \times 10^9 \text{ N}$

This is the tension of the two cables, so each cable would have ½ of this value.

The tension in a single cable is 12.5×10^9 N.

Again, apply the angle of 30° to find the minimum tension T_0 where $T_0 = T\cos\theta$.

$$T_0 = T\cos\theta = 25 \times 10^9 \text{ N} \cos 30° = 25 \times 10^9 \text{ N} (0.866) = 22.0 \times 10^9 \text{ N}$$

This is the tension in the pair of cables, hence the value of T for each cable is ½ that shown above; i.e. 11.0×10^9 N.

The equation describing the tension is: $T = T_0 \sqrt{1 + \dfrac{w^2}{T_0^2} x^2}$

Adding the values from above:

The equation describing the tension in both cables is:

$$T = 22.0 \times 10^9 \text{ N} \sqrt{1 + \dfrac{250 \times 10^6 \text{ N/m}}{22.0 \times 10^9 \text{ N}} x^2}$$

This can be approximated for a single cable as:

$$T = 11.0 \times 10^9 \text{ N} \sqrt{1 + \dfrac{125 \times 10^6 \text{ N/m}}{11.0 \times 10^9 \text{ N}} x^2}$$

The shape of the cable is: $y = \dfrac{1}{2} \dfrac{w}{T_0} x^2$

Adding the values from above (using ½ of the load for each cable):

$$y = \dfrac{1}{2}\left(\dfrac{250 \times 10^6 \text{ N/m}}{22.0 \times 10^9 \text{ N}} \right) x^2 = \dfrac{1}{2}\left(1.15 \times 10^{-2}/\text{m}\right)x^2$$

The shape of the cable is: $y = (5.77 \times 10^{-3}/\text{m})x^2$

Concept Reinforcement

1. The weight of a block of cast iron is 600 N. The block is on a surface at an angle of 20°, which is also cast iron. The coefficient of static friction between two cast iron surfaces is 1.1 and for kinetic friction is 0.15. Determine the forces of static friction and kinetic friction.

2. A bridge with a uniform load (w) of 300 MN/m (300×10^6 newtons/meter) is suspended by two cables. For a section of the bridge that is 80 m long, define the equations that describe the tension in the cable and the shape of the cable. Assume the angle at the upper end is 45°.

Appendix

Mechanical Engineering Answer Key – Unit 1

Section 1.1

1. Science is an investigation or analytical approach focused on developing new understanding of natural phenomena. Science evolved from philosophical studies. Early philosophers tried to explain things they saw in the world. Philosophers began to develop new ways of asking and answering questions, which lead to the development of science and the scientific method. Engineering is a profession that applies scientific knowledge to benefit the world.

2. Engineering benefits from new scientific developments. Likewise, science also benefits from new engineering developments. Scientists discovered electricity. The engineers put this to work and created a stable source of transportable energy. This allowed scientists to do interesting experiments using more complex instruments, which require a stable energy source.

3. Science generates new knowledge. Engineering is the practical application of this knowledge solve problems in the real world.

4. Biology is the study of living things, plants, animals and humans. Ecology is the study of the ecosystem, the impact of man on the environment and searches for solutions to the challenges of pollution and waste on Earth. Physics is the study of nature with the goal being to explain the reasons that things behave the way they do. Along with Newtonian mechanics and quantum mechanics, physics also covers thermodynamics, electromagnetism and the new theories of relativity.

   Mechanical engineering applies the principles of mechanics, which is a branch of physical science dealing with mechanical interactions between bodies and the interactions of forces and motion. Statics is a branch of mechanics focused on the analysis of the various forces active in rigid structure when it is in equilibrium. Dynamics is a branch of mechanics focused on the analysis of the interrelation between forces and motion of bodies.

   Civil engineering focuses on the structures needed for modern society, which includes buildings, bridges, roads, highways and large infrastructure projects like airports and hydroelectric facilities.

   Genetic engineering uses the principles of genetics and biology in combination with various engineering disciplines. The engineering disciplines include electrical, mechanical and hydrodynamic engineering.

Section 1.2

1. The SI base units have been defined as the standard by which all measurements can be compared. The SI base units, along with the entire system, were developed in 1960 to provide a uniform system of measures. Before that time, two systems had developed: the meter-kilogram-second system and the centimeter-gram-second. These were two systems that developed to reduce the variation in measures. The SI developed to further reduce the variation in measures internationally.

2.

Physical Quantity	Unit	Symbol
Length	meter	m
Time	second	s
Mass	kilogram	kg
Temperature	Kelvin	K
Amount of Substance	mole	mol
Electric Current	ampere	A
Luminous intensity	candela	cd

3. To convert between the SI units (meters) and the US Customary units: 1 foot (ft) = 0.3048 m (meters). Therefore, to determine the number of meters given the number of feet we multiply the number of feet × 0.3048.

    (21 feet)(0.3048 m/ft) = 6.4 m.
    Converting meters to kilometers: 1 m = 0.001 km
    Therefore 6.4 m = 0.0064 km = 6.4×10^{-3} km
    Converting meters to centimeters: 1 m = 100 cm
    Therefore 6.4 m = 640 cm = 6.4×10^2 cm
    Converting meters to micrometers: 1 m = 1,000,000 μm
    Therefore 6.4 m = 6,400,000 μm = 6.4×10^6 μm

4. To convert between the SI units (meters) and the US Customary units: 1 foot (ft) = 0.3048 m (meters). Therefore, to determine the number of meters given the number of feet we multiply the number of feet × 0.3048.

    (300 feet)(0.3048 m/ft) = 91.44 m.
    Converting meters to kilometers: 1 m = 0.001 km
    Therefore 91.44 m = 0.09144 km = 9.144×10^{-2} km
    Converting meters to centimeters: 1 m = 100 cm
    Therefore 91.44 m = 9,144 cm = 9.144×10^3 cm
    Converting meters to micrometers: 1 m = 1,000,000 μm
    Therefore 91.44 m = 91,440,000 μm = 9.1444×10^7 μm

Section 1.3

1. A scalar expresses a magnitude of the physical measurement with no direction. A vector expresses a magnitude of the physical measurement but also includes a direction.

2. When two vectors are in the same direction the addition of the two vectors is simple addition. Hence, we have 15 west + 5 west = (15 + 5) west = 20 west.

3. When two vectors are in opposite directions but the operation is subtraction this can be modified and expressed as an addition of the two vectors in the same direction. 10 north – 2 south = 10 north + (−2 south) = 10 north + (+2 north) = 12 north

4. When two vectors are in orthogonal directions (perpendicular) and the operation is addition this can apply the Pythagorean theorem to determine both the magnitude and direction. Graphically, the vectors can be added by drawing the first vector to scale and then starting the second vector from the end of the first. The resultant vector is a vector drawn from the origin of the first vector to the end of the second vector. Hence, this appears as a right triangle with the two original vectors as the base sides of the triangle and the resultant vector as the hypotenuse.

   For the magnitude: Resultant = $\sqrt{(5^2 + 10^2)}$ = $\sqrt{(125)}$ = 11.18 units

   For the angle: tangent = (opposite side)/(adjacent side) = 5/10 = 0.5/

   To determine the angle we obtain the arc tangent (= $\tan^{-1}$). $\tan^{-1}$ (0.5) = 26.5 degrees north west.

Section 1.4

1. Force is the term that describes the effort to push it or pull an object, or the amount of effort required to move an object over a distance. Something gets pushed or pulled and it moves.

2. If the velocity of the object changes in any way, either in magnitude or in direction, then an acceleration is experienced by the object.

3. Because both the amount and the direction of the push or pull carry relevant information, this infers that the force is directional, which means force is a vector quantity, with magnitude (amount) and direction.

4. • The object is drawn as a simple geometric shape, usually a box
   • All forces are shown as arrows
   • The forces are shown acting on the outer edge of the object
   • The origin of each force is the geometric center of the object
   • The length of the arrow shows the relative magnitude of the forces
   • The direction of the arrow shows the direction of the force
   • Each force vector is labeled

5. a. $F = (10 \text{ kg})(30 \text{ m/s}^2) = 300 \text{ kg·m/s}^2 = 300 \text{ N}$
   b. $F = (52 \text{ kg})(75 \text{ m/s}^2) = 3{,}900 \text{ kg} \times \text{m/s}^2 = 3.9 \text{ kN}$

6. a. An object is on a flat plane. It has a force of gravity = 980 N, a normal force which opposes the force of gravity and a force from the left of 100 N.

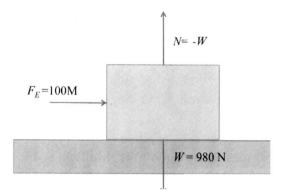

b. An object is on a ramp with an incline. It has a force of gravity = 1,000 N, there is a normal force which opposes the fraction of the force of gravity, a force down the ramp which is a fraction of the force of gravity and a force of friction of 100 N which is acting opposite the direction of movement. Define the magnitude of the forces as much as possible.

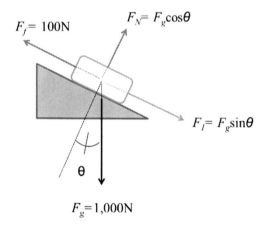

c. Create a free-body diagram of a person on a snowboard sliding down a hill

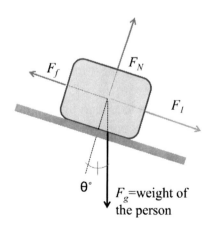

Section 1.5

1. "Moment," in mechanical engineering, is a term used to describe the effect on an object of a force that is acting on that object at a radial distance from the point of interest. The force acts in a way to bend the object around the point of interest. A force acts in a perpendicular direction at a distance from the point. The moment (M) is a vector quantity.

2. Place your right hand on the table with the fingers pointing straight away from your body and your thumb pointing straight up at a 90° angle from the fingers. Point your fingers in the direction of the distance (d). Then curl your fingers in the direction of the force as if wrapping your fingers around a pole with the thumb pointing up along the pole. Here, the thumb points in the direction of the moment vector.

3. The concept and formula for moment is very similar to that of torque. Moment is a more general examination of the effects of a force applied at a distance. While sometimes this causes a rotation at the axis point, it usually just causes stress. Torque, on the other hand, is a description of the rotational effect of the force applied at a distance and is usually related to a rotation at the axis.

4. With the formula for the moment being $M = Fd$ the moment $M = (300 \text{ N})(2 \text{ m}) = 600 \text{ Nm}$

5. With the formula for the moment being $M = Fd$ the moment $M = (300 \text{ N})(2 \text{ m})(\cos 30) = 600 \text{ Nm} (\cos 30) = 600 \text{ Nm} (0.866) = 519.6 \text{ Nm}$

Section 1.6

1. Force is vector and, therefore, must be presented with both a magnitude and a direction. For instance, describing a force as simple 10 N is insufficient. No direction is defined.

2. A vector is presented with a magnitude and direction. A positive direction is opposite to a negative direction. The negative sign translates to the same force pointing in the opposite direction. This negative force terminology is applicable, for instance, when a force is applied to slow a vehicle, or decelerate a car.

3. Vector summation is different from standard algebraic summation. The resultant vector F_R is longer than either of the initial force vectors and the direction is different as well. To sum the vectors graphically the first vector is drawn to scale in the correct magnitude and direction. This begins at an origin and the arrow is as long as the magnitude of the vector. We take the second vector and create a parallel vector that starts at the end (arrow tip) of the first vector. This is continued with each subsequent vector until all the vectors are lined up end to end. Then the resultant vector is from the origin of the first vector to the end of the last.

4. Vector 1: $F_1 = 10$ N at 30°; Vector 1 is pointing northeast and both its x- and y-components are positive.

$F_1x = F_1\cos\theta; F_1y = F_1\sin\theta$

$F_1x = F_1\cos\theta = 10 \text{ N}(0.866) = 8.66 \text{ N}$

$F_1y = F_1\sin\theta = 10 \text{ N}(0.5) = 5.0 \text{ N}$

Vector 2: $F_2 = 20$ N at 45°; Vector 2 is also pointing northeast and both its x- and y-components are positive.

$F_2x = F_2\cos\theta; F_2y = F_2\sin\theta$

$F_2x = F_2\cos\theta = 20 \text{ N}(0.707) = 14.14 \text{ N}$

$F_2y = F_2\sin\theta = 20 \text{ N}(0.707) = 14.14 \text{ N}$

Vector 3: $F_3 = 50$ N at 135°; Vector 3 is pointing northwest and therefore its x-component is negative while its y-component is positive.

$F_3x = F_3\cos\theta; F_3y = F_3\sin\theta$

$F_3x = F_3\cos\theta = 50 \text{ N}(-0.707) = -35.35 \text{ N}$

$F_3y = F_3\sin\theta = 50 \text{ N}(0.707) = 35.35 \text{ N}$

Resultant vector:

Total of the x-components: 8.66 N + 14.14 N + (−35.35 N) = −12.5 5 N

Total of the y-components: 5.0 N + 14.14 N + 35.35 N = 54.50 N

Magnitude of the resultant vector: $F_R = \sqrt{((-12.55)2 + (54.5)2)} = \sqrt{(3127.5)} = 55.9 \text{ N}$

To determine the direction we note that the x-component is negative and therefore the resultant vector must be pointing to the northwest.

The tangent of the angle = 54.5/−12.55 = −4.34

The arc tangent $(\tan^{-1}) = \tan^{-1}(-4.34) = -77°$

To define this angle using 0° as directly east this is 180 − 77 = 102.9°

The Resultant force vector is: 55.9 N at 102.9°

Section 1.7

1. The typical method to present a vector in three-dimensional space is to assume that its origin is at the position (0, 0, 0) and to stated its end point as a position in 3-D space by defining its coordinates in (x, y, z). Therefore the vector 3, 3, 3 would be a vector that begins at the point (0, 0, 0) and extends out to the position $x = 3$, $y = 3$ and $z = 3$.

2. To add the three vectors we assume that all coordinates are defined in the order x, y and then z.

$x = 3 + 2 + 1 = 6$ $y = 6 + 6 + 2 = 14$ $z = 9 + 8 + 3 = 20$

The resultant vector is defined as (6, 14, 20)

3. This is a way to present the resultant vector as a summation of the three components as defined by the basis vector. This is simply the coordinate with the basis vector terminology; i.e. $V = 7i + 2j + 9k$

4. In the basis vector all axes are orthogonal; i.e. they are all at 90° from each other. The x-axis is perpendicular to the y-axis and the z-axis. The y-axis is perpendicular to the x-axis and the z-axis. The z-axis is perpendicular to the y-axis and the x-axis.

   The basis vector is defined as:
   i is the vector from (0, 0, 0) to (1, 0, 0)
   j is the vector from (0, 0, 0) to (0, 1, 0)
   k is the vector from (0, 0, 0) to (0, 0, 1)

   Magnitude of the x-component: $F_x = F\cos\alpha$
   Where:
   F_x = the vector along the x-axis of with magnitude from point A to point x_1.
   α = the angle from the x-axis to the force vector.

   Magnitude of the y-component: $F_y = F\cos\beta$
   Where:
   F_y = the vector along the y-axis of with magnitude from point A to point y_1.
   β = the angle from the y-axis to the force vector.

   Magnitude of the z-component: $F_z = F\cos\gamma$
   Where:
   F_z = the vector along the z-axis of with magnitude from point A to point z_1.
   γ = the angle from the z-axis to the force vector.
   The angles α, β, and γ are all equal in the basis vector.

Section 1.8

1. Place your right hand on the table with the fingers pointing straight away from your body and your thumb pointing straight up at a 90° angle from the fingers. Point your fingers in the direction of the distance (d). Then curl your fingers in the direction of the force as if wrapping your fingers around a pole with the thumb pointing up along the pole. At this point, the thumb points in the direction of the moment vector.

2. To simplify the analysis at this stage all objects are assumed to be rigid bodies. This means that there is no bending of the object and the effects of the forces and moments are felt equally throughout the object. Contrast this to an object that is flexible or pliable. A force experienced at one end may not be experienced with the same intensity at the other end. This would complicate the analysis and, of course, is required at more advanced stages of mechanical engineering work. For now all objects will only be considered rigid bodies.

3. All forces are vectors and act along a direction called the line of action. In the illustration, the force is shown in red and the line of action is shown as a dashed line. The object experiences the force as if it is acting anywhere along the line of action.

4. First determine the cos60° which is 0.5.
Using the component of the force that is perpendicular to the radial distance:
$F\cos60° = 300 \text{ N}(0.5) = 150 \text{ N}$.
Moment $= (F\cos\theta)r = Fr\cos\theta = 150(0.7 \text{ m}) = 105.0 \text{ Nm}$
Using the distance (d) that is perpendicular to the line of action of the force,
$d = r\cos60° = 0.7 \text{ m}(0.5) = 0.35 \text{ m}$.
Moment $= Fd = F(r\cos\theta) = Fr\cos\theta = 300 \text{ N}(0.35 \text{ m}) = 105.0 \text{ Nm}$
In both cases, the moment is perpendicular to the plane described by the line of action of the force and the radial distance. The direction is counterclockwise.

Section 1.9

1. A couple is a pair of forces that are parallel, but in opposite directions. This pair of forces creates a moment about a point midway between the parallel lines of action of the forces.

2. Each force is equal and each is $r/2$ from the center point and therefore the two moments
$M_1 = F_1\dfrac{r}{2}$ and $M_2 = F_2\dfrac{r}{2}$ result in a single moment of: $M_o = F_1\dfrac{r}{2} + F_2\dfrac{r}{2} = 2F\dfrac{r}{2} = Fr$

3. In all cases the direction will be defined by the right hand rule, perpendicular to the plane that is defined by the two force vectors that create the couple.

Section 1.10

1. When the resultant forces of two systems are equivalent and the resultant moments are equivalent then the two systems are equivalent. This can be approached in two ways; either we can examine two random systems to determine if they are equivalent or we can change a known system to an equivalent system.

2. A force includes a magnitude and a direction. The force acts along a specific linear track, called its line of action. The line of action is perpendicular to the radial position to the point of interest, around which the moment is measured. A parallel force that is at a further radial distance is not acting along the same line of action. Along this line of action the force can move without changing the moment about the point of interest.

3. When moving a force off its line of action, a couple must be added to the new force system in the new location to compensate for the change. Here the original force system includes a force acting on one point, point A. The objective is to produce an equivalent force system with an equivalent force at a second point at a distance from point A, which is called point B.

An intermediate stage includes a couple which is created by adding a positive force and a negative force at point B both parallel to and the same magnitude as the original force at point A.

The final stage changes the couple (F_1 at point A and $-F_1$ at point B) to a moment (M_B) at point B. This is the moment that would be experienced at point B if the force had stayed at point A.

Section 1.11

1. A linear equation will have variables but no variables in the denominator of a fraction and no variables with a power higher than 1. The graph of a linear equation is a straight line. When there are two straight lines in a plane that intersect, the point of intersection is a coordinate (x, y), which is in the set of solutions for both of these lines. This can be extended to solve three, four and more simultaneous lines. This intersection point is the solution for both equations and hence is the solution for the "set" of these two linear equations.

2. The graphical method for solving the problem of the unique (x, y) coordinate that works for both lines is simply drawing each line on a graph and evaluating where they intersect. The intersection is the point at which the (x, y) coordinate is in the set of solutions for both of the lines.

3. When solving for the variables in a set of equations the primary goal is to eliminate one of the variables at a time. This can be done by manipulating one or both equations, and then using addition (or subtraction) of the two equations to remove one of the variables. This results in a single equation with a single unknown variable which can then be easily solved for that variable. Now that this variable is known it can be plugged into either of the initial equations to solve for the other variable.

4. When solving for the variables in a set of equations, the primary goal is to eliminate one of the variables at a time. This can also be done by using one of the equations to develop a substitution, where one of the variables is defined in terms of the other variable and a constant. This is used to replace the variable in the other equation so that there is only a single equation with a single unknown which is easily solvable. Now that this variable is known it can be plugged into either of the initial equations to solve for the other variable.

Section 1.12

1. Mechanical equilibrium is a condition of an object where the sum of the forces acting on the object is zero and the sum of the moments acting on the object is zero.

1. Mechanical Equilibrium: $\sum F = 0$ and $\sum M = 0$.

    When considering two-dimensional equilibrium we have:

    Mechanical Equilibrium: $\sum F_x = 0$, $\sum F_y = 0$ and $\sum M_z = 0$

The vector sum of the external forces and force components in the x-direction is zero and the sum of the external forces and force components in the y-direction is zero. The sum of the moments acting on the object in the z-direction is zero. All moments in a two-dimensional (x, y) system act in the z-direction. Therefore the object is in equilibrium and is not moving relative to the external forces and moments. It could also be that everything is moving at the same velocity. Because the whole object is in equilibrium this infers that a portion of the object is in equilibrium as well and can be analyzed separately.

2. In order to maintain a consistent treatment for forces and moments, the angles will be defined from the horizontal vector pointing to the right. This is established as zero degrees (0°) and the magnitude of the angle increases in the counterclockwise direction until the angle of the vector is pointing again directly to the right. At this point the vector has traversed one full rotation and is at an angle of 360° which is equivalent to 0°.

3. **Collinear Force System:** In a collinear force system all the forces are acting on the exact same line of action. With a multitude of forces all acting along the same line of action the forces must sum to zero for equilibrium to exist.

   Parallel Force System: In a parallel force system, all the forces act on parallel lines of action. With a multitude of forces all acting along parallel lines of action, the forces must sum to zero and the moments must sum to zero for equilibrium to exist.

   Concurrent Force System: In a concurrent force system all the forces are acting from the exact same origin. With a multitude of forces all acting along the same line of action the forces must sum to zero for equilibrium to exist.

   General Force System: In a general force system, all the forces are acting in random directions on the object and produce various moments. With a multitude of forces all acting along different lines of action the vector sum of the forces and force components must equal zero for equilibrium to exist. Also, in this case the moments about the z-axis must sum to zero for equilibrium to exist.

Section 1.13

1. Mechanical equilibrium is a condition of an object where the sum of the forces acting on the object is zero and the sum of the moments acting on the object is zero.

   Mechanical Equilibrium: $\sum F = 0$ and $\sum M = 0$.

   When considering three-dimensional equilibrium we have a more complicated set of requirements for equilibrium to exist:

   Mechanical Equilibrium of forces: $\sum F_x = 0$, $\sum F_y = 0$ and $\sum F_z = 0$.

   Mechanical Equilibrium of moments: $\sum M_x = 0$, $\sum M_y = 0$ and $\sum M_z = 0$.

   The vector sum of the external forces and force components is zero in the x-direction, in the y-direction and in the z-direction. Forces in a three dimensional (x, y, z) system act in any conceivable direction but can be deconstructed and presented as x, y and z components of the force.

The sum of the moments acting on the object is zero in the *x*-direction, in the *y*-direction and in the *z*-direction. Moments in a three dimensional (*x, y, z*) system act in any conceivable direction but can be deconstructed and presented as *x, y* and *z* components of the moment.

2. In order to maintain a consistent treatment for forces and moments, the angles will be defined from the horizontal vector pointing in the positive direction along an axis. This is established as zero degrees (0°) and the magnitude of the angle increases in the counterclockwise direction until the angle of the vector is pointing again directly to the right. In a 3D Cartesian coordinate system, the angle between vectors is what is important when evaluating force and moment components.

3. **Force System Concurrent at a Point:** In a force system that is concurrent at a point all the forces are acting from the exact same origin. With a multitude of forces all acting along the same line of action, the forces must sum to zero for equilibrium to exist.

   Force System Concurrent at a Line: In a force system that is concurrent at a line all the forces are acting on the exact same axis in the object. With a multitude of forces all acting along the same line of action the forces must sum to zero for equilibrium to exist.

   Parallel Force System: In a parallel force system all the forces act on parallel lines of action. With a multitude of forces all acting along parallel lines of action the forces must sum to zero and the moments must sum to zero for equilibrium

   General Force System: In a general force system all the forces are acting in random directions on the object and various moments are produced. With a multitude of forces all acting along different lines of action the vector sum of the forces and force components must equal zero for equilibrium to exist. Also, in this case the moments about the *x, y* and *z*-axes must sum to zero for equilibrium to exist.

Section 1.14

1. A truss is a support structure used to create mechanical stability. The most basic form of a truss is three support beams connected in a triangular shape. Under moderate levels of stress, a triangle will not change shape as easily as other simple geometric shapes like a square. Therefore, the forces that act on a truss will produce strain and moments in the beams, but the overall structure will not be seriously affected.

2. There are two major families of truss shapes, which are described by the outer shape of the overall structure rather than the inner webbing within the structure. The triangular truss is a very common shape in residential homes and many other applications. The flat shape is typical of the structure used in floors and flat construction.

3. Under moderate levels of stress, a triangle will not change shape as easily as other simple geometric shapes like a square. Therefore, the forces that act on a truss will produce strain and moments in the beams, but the overall structure will not be seriously affected.

4. The top side of the truss bears the load of rain, snow and wind plus the weight of the structure itself. Therefore, the top beams are in compression as the load presses on the beams.

The smaller webbing beams that rise up and connect to the middle of the top members are also experiencing the load and compression.

While the load presses down from above, the bottom beams are being pulled from both ends, experiencing tension. This is the same for the two longer webbing sections that rise to meet at the peak. They are stretched and therefore experience tension as well.

Section 1.15

1. This is the formula for force where $F = ma$. The Force $(F) = (100 \text{ kg})(9.8 \text{ m/s}^2) = 980 \text{ kg} \times \text{m/s}^2 = 980 \text{ N}$

2. These two vectors make up the two components of the resultant vector. The magnitude of the resultant vector is $F_R = \sqrt{((173)2 + (100)2)} = 200$

The angle from the horizontal is found by taking the arc tangent of $(100/173) = \tan^{-1}(0.578) = 30°$

Therefore the resultant vector is: 200 N at 30° Northeast

3.

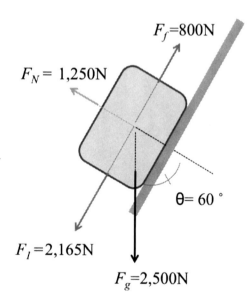

The normal force as a fraction of the force of gravity (2,500 N) is $F_g\cos60$

$F_N = F_g\cos60 = 2,500 \text{ N}(0.5) = 1,250 \text{ N}$

The component of the gravitational force that is acting on the block down the ramp is $F_g\sin60$.

$F_1 = F_g\sin60 = 2,500 \text{ N}(0.866) = 2,165 \text{ N}$

4.

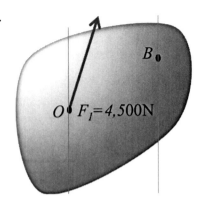

This is moving a force off its line of action. Therefore the force that is at the new location is a parallel force with the same magnitude and direction.

F_B = 4,500 N parallel to the original F_1.

The moment that is necessary to maintain equilibrium conditions with this move to point B is a moment that is defined by the force and the distance of separation. $M = Fr$.

$M = Fr$ = 4,500 N(0.35 m) =1,575 Nm

Mechanical Engineering Answer Key – Unit 2

Section 2.1

1. For the purposes of simplifying mechanical force analysis of trusses, a concept called an ideal truss is defined. In an ideal truss:

   • all sections are weightless; i.e. the weight of the sections do not impact the equilibrium calculations
   • all sections lie in a two dimensional plane; i.e. this is only a two dimensional problem
   • all sections are connected by smooth pins; i.e. the connections themselves do not change the forces or moments
   • all sections form triangular substructures within the main structure
   • all eternal loads are applied only at the joints

   The Rigid Body Assumption is also held. This states that the sections and the entire truss acts as a rigid body (that it does not twist or bend).

2. To analyze the forces acting on an object the standard procedure is to create a free-body diagram with all the forces acting from the geometric center of the object. Ideally all the moments would also be translated to the geometric center of the object as well.

   For the mechanical analysis of the forces acting on a truss there are two primary procedures:
   1. The Method of Joints and 2. The Method of Sections.

The Method of Joints applies the concepts of an Ideal Truss, assumes all forces act at a joint (a connection of two or more sections), and assumes all moments are the result of forces acting at joints. The objective of the Method of Joints is to define the forces and moments acting at the joints in question. This assumes that the forces are in balance and that the truss is in mechanical equilibrium; i.e. all forces in all directions sum to zero and all moments in all directions also sum to zero.

The Method of Sections also applies the concepts of an Ideal Truss, assumes all forces act at a joint (a connection of two or more sections), and assumes all moments are the result of forces acting at joints. The objective of the Method of Sections is to define the force's action on the various sections of the Truss. Whether the section in compression or is it in tension is determined through the Method of Sections.

3. Mechanical equilibrium is a condition of an object where the sum of the forces acting on the object and the sum of the moments acting on the object are both zero. In summary, this appears as mechanical equilibrium: $\sum F = 0$ and $\sum M = 0$. When considering two-dimensional equilibrium, we have:

Mechanical Equilibrium: $\sum F_x = 0$, $\sum F_y = 0$ and $\sum M_z = 0$.

Where:
F_x = a force or force component acting the x-direction in newtons (N)
F_y = a force or force component acting the y-direction in newtons (N)
M_z = a moment due to the forces acting in the x-y plane and acting the z-direction in newton-meters (Nm)

The vector sum of the external forces and force components in the x-direction is zero and the sum of the external forces and force components in the y-direction is zero. The sum of the moments acting on the object in the z-direction is zero. All moments in a two-dimensional (x, y) system act in the z-direction.

4. In order to maintain a consistent treatment for forces and moments, the angles will be defined from the horizontal vector pointing to the right. This is established as zero degrees (0°) and the magnitude of the angle increases in the counterclockwise direction until the angle of the vector is pointing again directly to the right. At this point the vector has traversed one full rotation and is at an angle of 360° which is equivalent to 0°.

Section 2.2

1. For the mechanical analysis of the forces acting on a truss there are two primary procedures; 1. The Method of Joints and 2. The Method of Sections.

The Method of Joints applies the concepts of an Ideal Truss, assumes all forces act at a joint (a connection of two or more sections), and assumes all moments are the result of forces acting at joints. The objective of the Method of Joints is to define the forces and moments acting at the joints in question. This assumes that the forces are in balance and that the truss is in mechanical equilibrium; i.e. all forces in all directions sum to zero and all moments in all directions also sum to zero.

The Method of Sections also applies the concepts of an Ideal Truss, assumes all forces act at a joint (a connection of two or more sections), and assumes all moments are the result of forces acting at joints. The objective of the Method of Sections is to define the force's action on the various sections of the Truss. Whether the section in compression or is it in tension is determined through the Method of Sections.

2. In the method of sections a line is draw through the truss across several members. This line can be a straight line or a curved line. The number of members "cut" can be many. After the truss has been "divided" a free body diagram is made, the x and y components of the forces are resolved and the equilibrium equations solved.

3. The method of joints starts with the boarder examination of the external forces, defining the equilibrium conditions of the whole truss with respect to the external forces. Then the method of joints examines each joint separately, starting from the external forces and produces a balanced equilibrium condition at each joint in turn until all the forces acting on all the joints in identified.

   In the method of sections, equilibrium conditions are assumed at each fractional section of the truss. Through this assumption the forces and moments are solved without solving all the forces and moments in adjoining joints and members. In this way the forces and moments can be resolved at a section within the truss without first solving all the forces one at a time starting from the external forces at the end supports (as in the method of joints). The portion of the truss from the external forces all the way to the members of the section of interest is considered a single object in equilibrium. Therefore the forces internal to this portion are not considered in the analysis.

Section 2.3

1. Within mechanical engineering, statics focuses on the analysis of external forces and moments on structures that are in static equilibrium. In a structure (a physical system) in static equilibrium, all the sections, components, members, etc. do not move relative to each other over time. Either the entire system does not move over time or all of it moves at the same velocity over time. Either way, the relative positions do not change and the way the subsections experience the external forces does not change over time.

2. Applying mechanical analysis of a structure to problems of statics initially involves several key assumptions:
   • the structure and the substructures are assumed to act as a rigid body (i.e. they do not deform),
   • the connecting joints are all smooth frictionless pins (i.e. they do not add or detract from the externally applied forces or moments)
   • the individual sections are weightless (i.e. the loads are only externally applied at the specific points of application of the external forces).

   These assumptions are used in this early mechanical analysis so that major mechanical principles can be the focus of the analysis. In more advanced mechanical analysis, these assumptions will be eliminated one by one.

3. The structural members of a mechanical system experience forces in two primary ways: only at the ends or in more than just at the ends.

An Axial Member is one that experiences the forces only at the ends and the forces can be summed so that the line-of-action is along the member, which will show that the member is only in tension or in compression.

A Non-Axial Member is one that also experiences external forces in the center span of the member and hence the line-of-action of the forces are not only along the length of the member. In this case, the resultant forces (the sum of the forces) are shown to have a line-of-action along the member and a line-of-action that is perpendicular to the member.

4. There are two general types of structural support: a pinned joint and a joint that is resting on a roller. The pinned joint can experience forces in both the x and the y directions and be in either the positive or negative direction. Supports that rest on a roller can only experience compression in the single direction. The compression is directed into the support. Because the roller allows movement in the other direction, these joints experience forces in one direction only and do not experience moments.

5. The three equilibrium equations for a two-dimensional system are:

A. $\sum F_x = 0$ (the sum of the forces in the x-direction is zero)

B. $\sum F_y = 0$ (the sum of the forces in the y-direction is zero)

C. $\sum M_z = 0$ (the sum of the moments about a point (in the z-direction) is zero)

Section 2.4

1. The center of mass in a two dimensional beam is the location along the beam where ½ the total mass is to the left and ½ the total mass is to the right. If several point masses are along a line the center of mass is a point in space at which ½ the total mass is to the left and ½ the total mass is to the right.

2. To determine the center of mass of a set of masses in a line the first step is to determine the force associated with each mass. A force is caused by a mass that is accelerated and a weight on Earth is caused by a mass accelerated by gravity. Once these masses are resolved into forces the moments that they create about a point are determined. Each force is at a different distance from this point and therefore creates a different moment. These moments are summed and then divided by the total force. The result is a location at a distance from the point at which the same total moment would be created if a force that is equal to the total force were applied. To determine the actual total mass requires only that the total force be divided by the acceleration of gravity.

3. To determine the center of mass of a set of forces in a line the moments of the various forces (a force is caused by a mass that is accelerated by gravity) about a reference point are determined. Each force is at a different distance from this point and therefore creates a different moment. These moments are summed and then divided by the total force. The

result is a location at a distance from the reference point at which the same total moment would be created if a force that is equal to the total force were applied. To determine the actual total mass requires only that the total force be divided by the acceleration of gravity.

4. The centroid of area is the same concept as the center of mass, but applied to a two-dimensional area rather than a single dimension; i.e. a line. This is also commonly called the center of gravity or center of mass of the area. This could be considered the balance point where the area would be able to balance on a point without tipping in any direction.

5. If the object is geometric and the sides can be bisected, then the first step is to draw a line bisecting each of the various sides. The centroid of the area will be at the intersection of these lines. If the object is not geometric, i.e. it is not symmetrical, then graphical methods are much more difficult.

6. The centroid of an area is calculated by dividing the area in one dimension into many parallel sections. Then the centroid of each area is established, keeping in mind that the center of mass should be the geometric center of the thin section. These sections are then added together and divided by the total area. This is done in a similar way to the calculation for the center of mass, which is the summation of the various centers of mass all divided by the total mass. Since the mass in an area is uniform, the sum of the area is equivalent to the sum of the mass (or weight) for each section and for the total area. Note that because the mass density is uniform, mass is removed from the equations for centroids of area.

Section 2.5

1. The centroid of an area is determined along two axes. First, to determine the x-coordinate of the centroid of the area the area is divided up into uniform sections and the moments about the y-axis are determined. These moments are summed and then divided by the total area. This results in a line parallel to the y-axis at some specific x-coordinate. This line can be considered the x-coordinate of the centroid of the area. It is a line which bisects the area with ½ the total area to the right and ½ the total area to the left. Second, to determine the y-coordinate of the centroid of the area the area is divided up into uniform sections and the moments about the x-axis are determined. These moments are summed and then divided by the total area. This results in a line parallel to the x-axis at some specific y-coordinate. This line can be considered the y-coordinate of the centroid of the area. It is a line which bisects the area with ½ the total area above and ½ the total area below. Lastly, the centroid of the area is the point coordinate (x, y) which is the intersection of these two lines.

2. The centroid of the area is equivalent to the center of mass of the object or area with the mass density of the area is uniform. If the mass density is uniform then a fraction of the area equals the same fraction of the mass and if a line divides the area with ½ of the total area to the left and ½ of the total area to the right, this is equivalent to saying that ½ of the total mass is to the left and ½ of the total mass is to the right.

3. A composite area is a set of shapes, usually geometric, that are joined together to create a single grouping. For example, using several different sized squares, the grouping will not appear like a single uniform square but can still be treated as a single total area.

4. The centroid of the area of a composite area is, as with the centroid for a uniform area, the point coordinate (x, y) at which the area is bisected in both the x and the y directions; i.e. ½ the total area to the right and ½ the total area to the left; ½ the total area above and ½ the total area below.

5. The formula for the centroid of a composite area is similar to the formula for the center of mass of a set of point masses. The area moments are calculated in each direction (usually the x and y direction). An area moment is calculated as the area × distance from a reference point. These area moments are summed and then divided by the total area. This is done along the x-axis which results in a x-coordinate with ½ the total area to the left, ½ the total area to the right. This is done along the y-axis which results in y-coordinate with ½ the total area above and ½ the total area below. The centroid of the composite area is the point coordinate (x, y) which is the intersection of these two lines.

Section 2.6

1. A load is the term used for the weight associated with a mass above a mechanical system. When using early models for mechanical analysis, the load was a single force applied at a point to the system but it is possible for more than a single external load to be experienced by the system. These load are always shown as point loads. The units of a point load are always in the form of force or weight with the most common units being either newtons or pounds.

2. A distributed load is different from a point load. Instead of being applied at a distinct point as in point loads a distributed load is spread out over a length or even over an area. An example would be a block that rests on a mechanical system and exposes the system to a force that is spread out along the entire area of the block. While the block may have a total force of 1000 N that it applies to the mechanical system the fact that the load of the block is distributed means that no single point on the mechanical system actually experiences exactly 1000 N. If the contact between the block and the system is a total of 2 meters in length, then the load is spread over those 2 meters and the system experiences 500 N/m of length.

3. A uniform load is expressed as a force per unit distance and is consistent along the entire length of the exposure. For example, a uniform load would be expressed as 10 N/m or 400 lb/ft. A uniformly increasing distributed load is a linearly increasing load and the force per unit distance changes from point to point along the length that the system is exposed to the force. For example, a uniformly increasing and distributed load would have a load of 100 N/m at one end and a load of 1000 N/m at the other end. It would have a triangular shaped set of force arrows in a free body diagram.

1. The centroid is the same as the center of mass or the center of gravity. It is calculated using dimensions of length, rather than mass or force, because several assumptions made are made as part of these calculations.

   Assumption 1 is that the density, the amount of mass per unit volume is constant.

   Assumption 2 is that the thickness of the material is constant.

   Using these two assumptions, the calculation for centroid in a two-dimensional system requires only the use of unit area and distance from an axis. The center of mass requires the use of a mass or force unit in the calculations.

2. A three-dimensional system has four centroids: the x-axis centroid; y-axis centroid; z-axis centroid; and the overall volumetric centroid, which is located at the intersection of the x, y and z centroids.

   The centroid of a volume is a point within an object at which the 3-dimensional object will be balanced like a helium balloon in a 3-dimensional space. The centroid about the x-axis is similar but only applicable in the one dimension. The volume, if positioned on a line at the x-centroid, will be balanced in that dimension. For the y-centroid, the line will be perpendicular to the x-centroid line and the z-centroid line and the volume will again be balanced. For the z-centroid, the line will be perpendicular to both the x-centroid line and the y-centroid line and the volume will again be balanced.

3. The centroid of a volume is determined using three formulas. One calculates the centroid about the x-axis, the second about the y-axis and the third about the z-axis.

$$x_c = \frac{\sum_{i=1}^{n} x_i V_i}{\sum_{i=1}^{n} V_i} = \frac{1}{V_T}\sum_{i=1}^{n} x_i V_i \qquad y_c = \frac{\sum_{i=1}^{n} y_i V_i}{\sum_{i=1}^{n} V_i} = \frac{1}{V_T}\sum_{i=1}^{n} y_i V_i \qquad z_c = \frac{\sum_{i=1}^{n} z_i V_i}{\sum_{i=1}^{n} V_i} = \frac{1}{V_T}\sum_{i=1}^{n} z_i V_i$$

4. The centroid of a volume is determined using the three formulas for the three centroids about the 3 Cartesian coordinate axes. One for the centroid about the x-axis, the second about the y-axis and the third about the z-axis. Each formula takes the sum of the area moments about an axis and divides this sum by the total area. The result is a coordinate for either the x, y or z axis. To obtain the centroid of the volume the intersection point of the three centroids about the Cartesian axes is defined. This is the x-coordinate from the x-centroid, the y-coordinate from the y-centroid and the z-coordinate from the z-centroid.

5. The calculations take into consideration all three dimensions within a coordinate system. The calculation of the centroid about each axis is comparable to the calculation of the moment of the line about each axis.

For the x-coordinate of the centroid of a line: $x_c = \dfrac{\sum\limits_{i=1}^{n} x_i L_i}{\sum\limits_{i=1}^{n} L_i} = \dfrac{1}{L_T}\sum\limits_{i=1}^{n} x L_i$

This is the sum of the collection of each unit of length × distance of that unit length from the x-axis. This is then divided by the total length.

For the y-coordinate of the centroid of a length: $y_c = \dfrac{\sum\limits_{i=1}^{n} y_i L_i}{\sum\limits_{i=1}^{n} L_i} = \dfrac{1}{L_T}\sum\limits_{i=1}^{n} y L_i$

This is the sum of the collection of each unit of length × distance of that unit length from the y-axis. This is then divided by the total length.

For the z-coordinate of the centroid of a length: $z_c = \dfrac{\sum\limits_{i=1}^{n} z_i L_i}{\sum\limits_{i=1}^{n} L_i} = \dfrac{1}{L_T}\sum\limits_{i=1}^{n} z L_i$

Section 2.8

1. The centroid of a line is the point in 3-dimensional space on an oddly shaped 3 Dimensional line at which the distance of the line in each direction is balanced with an equal amount of distance in the x, y and z directions.

2. The centroid of an area is point in two-dimensional space which is at the intersection of the centroid about the x-axis and the centroid about the y-axis of an area.

3. The first Pappus-Guldinus theorem describes a simplified formula for determining the area of a surface of revolution using the centroid of the line about that axis. The line cannot intersect the axis of rotation. An example of a surface of revolution would be the outer surface of a soda can. Using a line in 3-dimensional space at a distance from an axis and rotating this line about an axis in the y-direction creates a barrel shape in the y-direction. The first Pappus-Guldinus theorem provides a simplified formula to determine the surface area of the outside of that "barrel" using the centroid of that line.

4. The second Pappus-Guldinus theorem describes a simplified formula for determining the volume of a volume of revolution using the centroid of the area about that axis. The area cannot intersect the axis of rotation. Using an area in 3-dimensional space that is defined by a central axis and an outside line and rotating this area about an axis in the y-direction creates a solid barrel shape in the y-direction. The second Pappus-Guldinus theorem provides a simplified formula to determine the volume occupied by that "barrel" using the centroid of that area.

5. The formula for the volume of the volume of revolution: $V = 2\pi \bar{y} A$

$V = 2\pi \bar{y} A = 2\pi(2 \text{ m})25 \text{ m}^2 = 314.16 \text{ m}^3$

Section 2.9

1. Along a line with several point forces (or masses) the moments of all of these forces about a reference point can be obtained. These force (or masses) can also be summed. The center of mass along a line is the point at which the total force (mass) of a set of point forces (masses) can be applied so that the moments about a reference point is the same as that for all the individual point forces (masses).

2. The center of mass of an area is the intersection of the center axis of the mass along the x-axis with the center axis of mass along the y-axis. This is the point in the area at which the mass on either side of line bisecting the area through this point leaves ½ of the mass of the object on either side of the line.

3. The center of mass of a volume is the intersection of the center axis of the mass along the x-axis with the center axis of mass along the y-axis and the center axis of mass along the z-axis. This is the point in the volume at which the mass on either side of plane bisecting the volume through this point leaves ½ of the mass of the object on either side of the plane.

4. The centroid of the area is the same as the center of mass for an area when the mass density of the area is uniform. They are both the point on the area through which a line bisecting the area will leave ½ of the area (or mass) on either side of the line. When the mass density of the area is not uniform the centroid of the area will probably not be at the same location as the center of mass.

5. The centroid of a volume is the same as the center of mass for the volume when the mass density of the volume is uniform. They are both the point in the volume through which a plane bisecting the volume will leave ½ of the volume (or mass) on either side of the plane. When the mass density of the volume is not uniform the centroid of the volume will probably not be at the same location as the center of mass.

Section 2.10

1. The center of mass of an area is the intersection of the center axis of the mass along the x-axis with the center axis of mass along the y-axis. This is the point in the area at which the mass on either side of line bisecting the area through this point leaves ½ of the mass of the object on either side of the line.

2. The centroid of the area is the same as the center of mass for an area when the mass density of the area is uniform. They are both the point on the area through which a line bisecting the area will leave ½ of the area (or mass) on either side of the line. When the mass density of the area is not uniform the centroid of the area will probably not be at the same location as the center of mass.

3. A composite area is a collection of separate shapes which are treated as a single unit. If the separate shapes are geometric then the centroids are relatively easy to determine using formulas. When they are not geometric shapes the determination of the centroid of the individual shapes requires advanced math.

4. The center of mass for a composite area is determined similar to the way the center of mass for a set of point masses on a line is determined. The moments are determined about a reference point and this total is divided by the total mass (of all three shapes). This results in a location along that axis at which a point mass would be applied which would results in the same total moment as that for the set of point masses at different distances. Here, mass and force are interchangeable because the difference is just the factor of gravity (9.8 m/s²). For a composite area this set of calculations is performed along the x-axis and then along the y-axis. These two coordinates form the x, y coordinates of the center of mass for the area.

5. These formulas create "mass" moments (rather than "force" moments) because they use mass terms in the formula for the moments instead of forces. Using the y-axis (as a reference point where $x = 0$) the mass moments are summed and then divided by the total mass. The same is performed using the x-axis (as a reference point where $y = 0$).

The calculation for the center of mass of an area about the y-axis: $x_c = \dfrac{\sum\limits_{i=1}^{n} x_i m_i}{\sum\limits_{i=1}^{n} m_i}$

Where:
x_i = the distance to the unit of mass (m_i) from the y-axis in meters (m)
m_i = the mass of the composite shapes in kilograms (kg)

For 3 areas the resulting formula is: $x_c = \dfrac{x_1 m_1 + x_3 m_3}{m_1 + m_3}$

The calculation for the center of mass about the x-axis: $y_c = \dfrac{\sum\limits_{i=1}^{n} y_i m_i}{\sum\limits_{i=1}^{n} m_i}$

Where:
y_i = the distance to the unit of mass (m_i) from the x-axis in meters (m)
m_i = the mass of the composite shapes in kilograms (kg)

For 3 areas the resulting formula is: $y_c = \dfrac{y_1 m_1 + y_3 m_3}{m_1 + m_3}$

Section 2.11

1. Inertia is a tendency. It is a measure of how difficult it is to change what an object is doing.

2. When a body is not moving it is inert (has inertia). It will not move without some force causing it to move. If a body is moving at a constant velocity, there is no change in velocity from one time to the next. Hence, there is no acceleration, and therefore no force acting on

the body. This moving body also has inertia. In order for it to accelerate, which by definition, is to change its velocity, a force must be applied to the body.

3. Linear inertia measures the resistance to change in a framework of straight-line motion. When an object makes a turn this is still considered to be examined in a straight-line framework because the acceleration and the inertia is examined based on straight-line directions. Rotational inertia is similar to linear inertia in that it is a measure of resistance to change. Rotational inertia considers the situation where all the world is turning at the same number of revolutions per second. What does it mean to "turn" in this situation when we are all revolving around an axis constantly? Looking at this from a straight-line framework a mass will only revolve around an axis if the centripetal force is constant and sufficient. Looking at this from a rotational framework the "moment of inertia" is a measure of the ability of an object to resist the rotational movement. This is similar to the way mass is a measure of the ability of an object to resist linear movement.

4. Linear kinetic energy uses the formula: $K = \frac{1}{2}mv^2$

   This is ½ × mass(m) × linear velocity squared (v^2)

   Angular kinetic energy uses the formula: $K = \frac{1}{2}I\omega^2$

   This is ½ × moment of inertia (I) × angular velocity squared ($\omega 2$)

   Here the moment of inertia is equivalent to the mass and the angular velocity is equivalent to the linear velocity. The units will both be in joules of energy.

Section 2.12

1. The moment of inertia is calculated as the mass × the square of the distance to the center of mass.

   Moment of Inertia: $I = mr^2$

   Where:
   I = the moment of inertia in kilogram-meters squared (kg-m^2)
   m = mass in kilograms (kg)
   r = radial distance from the axis of rotation to the center of mass in meters (m)

2. The moment of inertia is the measure of the resistance of an object to a change in movement in a rotational framework similar to the way mass is a measure of the resistance of an object to change in movement in a linear framework.

3. The formula for the moment of inertia of a solid sphere is: $I = \frac{2}{5}mr^2$

Where:

I = the moment of inertia in kilogram-meters squared (kg-m²)
m = mass in kilograms (kg)
r = radial distance from the axis of rotation to the center of mass in meters (m)

4. Using the formula: $I = \frac{2}{5}mr^2$

$$I = \frac{2}{5}mr^2 = \frac{2}{5}(300 \text{ kg})(0.320 \text{ m})^2 = 12.29 \text{ kg} \times \text{m}^2/\text{s}$$

Section 2.13

1. All objects exist in the Cartesian coordinate system, which includes x, y, and z axes. This can be presented in a variety of ways but there are two generally common presentations. One has the x-axis directed towards the lower right and the y-axis is towards the upper right and the z-axis is straight up.

2. There are times when the entire coordinate system is shifted by a consistent defined amount or rotated to a new orientation with respect to the object. Once the coordinate system is shifted, the moment of inertia about each of the axes in this new position will be different from that determined using the standard coordinate system.

3. A translated coordinate system is a coordinate system that is shifted linearly along either the x, y or z axis. A rotated coordinate system leaves the origin in the same location but the axes are shifted by an angular amount rather than a linear amount. The new axes are at a different angle from the original coordinate system.

4. With an original set of axes the moment of inertia may be at a magnitude that is incorrect (either too large or too small) for the application. If the axes are rotated the moment of inertia about the new axes cane be altered with respect to the original set of axes.

5. The Principal Axes occur when the coordinate system is rotated to a point where the moment of inertia is minimum about one of the axes. At this point, the moment of inertia is called the principal moment of Inertia.

Section 2.14

1. The center of mass of an object is the intersection of the center axis of the mass along the x-axis with the center axis of mass along the y-axis and the center axis of mass along the z-axis. This is the point in the object at which the mass on either side of a plane bisecting the object through this point leaves ½ of the mass of the object on either side of the plane.

2. Within the dimensions of an object, there is a position where the moment of inertia about each axis is at a minimum. This is the center of mass of the object about that specific axis. This will rarely be the minimum moment of inertia about all the axes at the same time, because the minimum moment of inertia about all axes at the same time would only occur with a spherical volume in 3 dimensions. Therefore, there will normally be a minimum moment of inertia in only one dimension about a specific axis. Because the mass, area or volume of the object does not change, this minimum moment of inertia will occur where the distances are at a minimum, somewhere near the center of the object.

3. The formula for the minimum moment of inertia is the moment of inertia at the center of mass of the object. At any position outside of the center of mass the moment of inertia will be larger.

4. The parallel axis theorem gives a formula for determining the moment of inertia about any parallel axis if the mass and the minimum moment of inertia are known. The minimum moment of inertia is the moment about an axis through the center of mass of the object.

Parallel Axis Theorem: $I_p = I_c + mr^2$

Where:
I_p = the moment of inertia about a parallel axis in kilogram-meter squared (kg-m2)
I_c = the minimum moment of inertia about an axis in kilogram-meter squared (kg-m2)
m = mass in kilograms (kg)
r = the distance from the axis with a minimum moment of inertia to the new parallel axis in meters (m).

5. Using the formula for the Parallel Axis Theorem: $I_p = I_c + mr^2$

$I_p = I_c + mr^2 = 1{,}000 \text{ kg·m}^2 + (300 \text{ kg})(2 \text{ m})^2$

$I_p = I_c + mr^2 = 1{,}000 \text{ kg·m}^2 + 1{,}200 \text{ kg·m}^2 = 2{,}200 \text{ kg·m}^2$

Section 2.15

1. A frame is a broad family of mechanical support structure that includes other structures such as trusses. Frames utilize axial and non-axial members, pinned joints and roller supports, etc.

2. The three equations of equilibrium in a two dimensional problem are:

$\sum F_x = 0$ (the sum of the forces in the x-direction is zero)

$\sum F_y = 0$ (the sum of the forces in the y-direction is zero)

$\sum M_z = 0$ (the sum of the moments about a point (in the z-direction) is zero)

3. The three equations of equilibrium in a two dimensional problem are:

$\sum F_x = 0$ (the sum of the forces in the x-direction is zero)

$\sum F_y = 0$ (the sum of the forces in the y-direction is zero)

$\sum F_z = 0$ (the sum of the forces in the z-direction is zero)

$\sum M = 0$ (the total of the moments about every axis is zero)

4. The centroid of the area is the same as the center of mass for an area when the mass density of the area is uniform. They are both the point on the area through which a line bisecting the area will leave ½ of the area (or mass) on either side of the line. When the mass density of the area is not uniform the centroid of the area will probably not be at the same location as the center of mass.

5. This is an application of the formula of the moment of inertia for a sphere: $I = \frac{2}{5}mr^2$

$$I = \frac{2}{5}mr^2 = \frac{2}{5}\left(400 \text{ kg}(0.75 \text{ m})^2\right) = \frac{2}{5}400 \text{ kg}\left(0.5625\text{m}^2\right) = 90 \text{ kgm}^2$$

Mechanical Engineering Answer Key — Unit 3

Section 3.1

1. Friction is the term used to describe the force of resistance to movement. Friction exists when one surface acts against another. For example, a paperclip will continue to hold sheets of paper together because of the friction between the clip and the paper. Likewise, if those sheets of papers are between the pages in a closed book, the friction between the sheets of paper and the pages of the book holds the sheets in the book.

2. Static friction occurs between two stationary objects. A pair of shoes on the floor, tools on a workbench, art paint tubes and brushes on an easel, and boxes of birthday presents on a table all exhibit static friction when they are not moved. They "adhere" to the surface on which they stand. Remember that once you get a stationary object moving, it seems to move with less effort. The force of your push exceeded the force of static friction. It required a stronger push to get it moving initially and that is the effect of static friction.

3. Kinetic friction is a continuous resistive force that acts between objects when they are moving against each other. Essentially, once the force applied to the object overcomes static friction, kinetic friction comes into play.

Using the formula for the force of static friction: $F\text{fric} = \mu_s N$

The magnitude of the Normal force is the same as the weight, 500 N. The direction is opposite. The normal force is $N = -500$ N.

$F\text{fric} = \mu_s N = (0.78)(-500 \text{ N}) = -390$ N. This is 390 N in the direction opposite the movement.

Therefore, this block requires 390 N of force applied horizontally to get it to move from a static position.

Ffric $= \mu_k N = (0.56)(-500 \text{ N}) = -280 \text{ N}.$

This is 280 N in the direction opposite the movement. Therefore, once this block is moving it continues to require 280 N of force applied horizontally to keep it moving.

Force of static friction: Ffric $= \mu_s N$

The magnitude of the Normal force is the same as the weight, but opposite in sign, $N = -600 \text{ N} \cos\theta.$

Ffric $= \mu_s N = (1.1)(-600 \text{ N})(\cos(30°)) = -660 \text{ N}(0.866) = -572 \text{ N}.$

Therefore, this block requires 572 N of force applied at an angle of 30° down the ramp to get it to move from a static position at this angle. Once the block is moving the kinetic friction is the resistive force.

Ffric $= \mu_k N = (0.15)(-600 \text{ N})(\cos(30°)) = -90 \text{ N}(0.866) = -77.9 \text{ N}.$

Therefore, this block requires 77.9 N of force applied at an angle of 30° down to keep it moving at this angle.

Section 3.2

1. A beam is the term that refers to sections of a support structure. They are normally straight sections, but there are reasons that they may be other than straight. Beams support a load. The strength of a beam relates to the amount the beam will bend when a particular load is placed on the beam. The shape of a beam influences the ability of the beam to withstand a load and resist bending due to the applied load. A typical beam has a length that is many times larger than the dimension of the height or width. When discussing the various common shapes of a beam, this generally refers to the cross section.

2. Most beams carry vertical loads; i.e. loads from above. In virtually all situations where beams are used, the objective is to transfer the load to vertical supports, such as columns, walls or other beams.

   Compressive Stress: When the load is from the top the upper side of the beam experiences compressive stress along the horizontal length of the beam. An exaggerated amount of deflection occurs when the top bends down under the force of the load. This causes the horizontal compressive stress on the upper ½ portion of the beam.

   Tensile Stress: The bottom side of the beam experiences tensile stress along the horizontal length of the beam. Under the force of the load, the bottom side of the beam extends out with the deflection. This is the cause of the tensile stress.

   Shear Stress: Because the load is centered on the beam and the supports are at the ends of the beam, there is a strong possibility that shear stress exists within the beam itself.

3. When the load is from the top the upper side of the beam experiences compressive stress along the horizontal length of the beam. An exaggerated amount of deflection occurs when the top bends down under the force of the load. This causes the horizontal compressive stress on the upper ½ portion of the beam. The bottom side of the beam experiences tensile stress along the horizontal length of the beam. Under the force of the load, the bottom side of the beam extends out with the deflection. This is the cause of the tensile stress.

4. When the load is centered on the beam and the supports are at the ends of the beam, there is a strong possibility that shear stress exists within the beam itself. If we divide the beam up into many vertical layers, we can more readily explain the shear stresses. Starting from the center and moving to the left, the central layers of the beam, where the load is applied, have relatively more force pushing the layers down compared to the next layer to the left. The same concept applies as we move from the center to the right. We see the same situation, with the center portions experiencing more downward force than those layers to the right.

5. Because the load is being supported by single-point structural supports, the beam experiences a moment. This is a bending moment, which is sometimes also called internal torque.

Section 3.3

1. Axial force is a force experienced by a structural member in the direction along the length of the member. Axial force is usually drawn as a tensile force, with the direction moving out of the member and normal to the cross section of the member. Axial force is the force experienced by an axial member, which is a member that experiences forces only at the ends. Using this convention, with the force always pointing out of the axial end, if the value of the force is positive, it is then a tensile force and if it is negative, it is then a compressive force.

2. Axial force is the force experienced by an axial member, which is a member that experiences forces only at the ends. Axial stress is measured in force per unit area. Axial stress is similar to the concept of pressure in a fluid. Both use the same base units of force per unit of cross sectional area.

3. Mechanical equilibrium is when the sum of the forces is zero. Forces are vectors and therefore, for equilibrium to exist, the x-components of the various force vectors must sum to zero. This requirement applies to the y-components of the various force vectors and the z-components of the various force vectors. Also, all moments must sum to zero as well. The moments will be in various directions depending upon the plane described by the distance and the force.

4. Within the structure of a beam, the material can be considered to exist in layers. When there is a concentrated force (load) acting on a beam at a point, the material at that point will tend to move in the direction of the force. The material layers around this point will resist movement and hence there will be a force differential. The layers furthest away from the point force move least. This differential is called transverse shear stress and is always parallel to the direction of the acting force.

5. If the beam is experiencing a load but is in equilibrium, this means that the beam is not moving relative to the outside forces. Within the material of the beam itself the layers of the material must experience shear forces due to the load. The amount of shear force experienced will be larger near the load and less at further distance from the load. To counter the load in each fractional component of the beam a bending moment must exist to maintain equilibrium – a balance of the effects of the forces and the moments on the fraction of the beam itself.

Section 3.4

1. Axial force is a force experienced by a structural member in the direction along the length of the member. Axial force is usually drawn as a tensile force, with the direction moving out of the member and normal to the cross section of the member. Axial force is the force experienced by an axial member, which is a member that experiences forces only at the ends. Using this convention, with the force always pointing out of the axial end, if the value of the force is positive, it is then a tensile force and if it is negative, it is then a compressive force.

2. A non-axial member is a structural section that experiences forces anywhere along the length of the member. These forces can be experienced at any angle. The result is a more complicated set of issues related to the forces, stresses, moments and deformation of the member.

3. Within the structure of a beam, the material can be considered to exist in layers. When there is a concentrated force (load) acting on a beam at a point, the material at that point will tend to move in the direction of the force. The material layers around this point will resist movement and hence there will be a force differential. The layers furthest away from the point force move least. This differential is called transverse shear stress and is always parallel to the direction of the acting force.

4. Bending moments (also called internal torque) will be experienced by all portions of the beam under equilibrium conditions due to the combination of the load pushing down and the supports pushing up.

Section 3.5

1. A distributed load is actually much more common than point loads and can be visualized as a block with mass resting on a structural member. The most common distributed load is the uniform load where a load of constant magnitude (per unit of length) is applied over the entire length of the beam. The mass of a block experiences the gravitational acceleration and this creates the weight of the object or load. This load is spread evenly along the beam for the full length of the block.

2. The free-body diagram for a distributed load no longer shows a single force acting from the center of the object. This can complicate the equilibrium calculations but there is a common procedure to return us to a force applied at a point. That is called resolving the distributed loads and entails finding the centroid of the load (the center of gravity) and determine the total force applied by the load.

Starting with uniformly distributed loads the centroid is easily found. For a rectangle or a uniform load "area" the centroid is the intersection of the two diagonal lines.

3. The centroid of a triangle is also a simple formula of the intersection of two lines, each at ⅓ from the wider part of the triangle, with one line parallel to one of the bases of the triangle.

4. The shear force is a fraction of the total load and is proportional to the load. When examined from the left end the shear force in this section is equal to the vertical support at point A plus the load/length × the length starting at the distributed load. This continues until the end of the distributed load.

5. The bending moment that the beam experiences through the length below the distributed load is also dependent upon the length and the load/length. It is related to the shear force, which is related to the load/length.

Section 3.6

1. Primarily used as a bridge support, structural cabling is more flexible than solid structural members. It does, however, have the same material properties in terms of the amount of stress it can withstand. A cable is a flexible structural support. In general, a cable is assumed to offer no resistance to bending. Therefore, a cable cannot transfer bending moments. In addition, because of this assumption, cables cannot transfer transverse loads and all loads are assumed to be represented as tensile forces along the cable (normal to the cross sectional area of the cable).

2. Cables can be constructed using several parallel wires braided together along the length of the wire. They can also be constructed with a solid core with many parallel wires wound around the core. The primary objective of this construction is to create a structural element that is more flexible than a beam while also having the tensile strength comparable to the other structural members. As the amount of individual wires increases, the tensile strength of the group of wires which are components of the cable increases.

3. **Galvanized Bridge Wire:** This is the basic wire used to produce parallel wire bridge cables. Various sizes are available, with a typical diameter of approximately 0.2 cm.

    Galvanized Bridge Strand: A bridge strand is constructed of many bridge wires twisted together to form a single support. The bridge wires may have many different diameters. The bridge strand may consist of a few or many individual bridge wires.

    Galvanized Bridge Rope: A bridge rope is constructed of several, typically six, bridge strands. These strands are usually braided around a center core strand.

4. **Linear Transmission (A Bowden cable):** Though the cable will have one or more turns along the path, the cable sheath will be stiffer in order to improve the transmission of a linear force. This can be a pushing force, which translates to compression on the cable from one end, transmitting a push to an element at the other end of the cable. This can also be a pulling force, which translates to tension on the cable from the operator end. This transmits a pull to the other end of the cable.

   Rotational Transmission: In this type of transmission, the cable is moved linearly into and out of a housing that translates this linear motion into a rotation of a wheel within the housing.

   Angular Motion: When there is a need for a structural member to be able to flex in position and also transmit a turning or twisting motion, the choice is often a type of cable that remains and includes several "elbow" joints so that the angular motion is possible.

5. **Wire rope cables:** Wire rope cables are a braided set of narrow gauge wire with no external wrapping around the braid. These are extremely flexible, which results in a very low ability to transmit tensile forces and no ability to transmit compressive forces. These typically transmit force with low efficiency.

   Helix cables: Helix cables consist of a straight core of wires wrapped by one or more layers of heavier wire. These are very flexible, but this flexibility results in a reduction in the amount of tensile or compressive forces that can be transmitted. Within the range of loads that are allowed, these cables are very efficient.

   Flat wrap cables: Flat wrap cables also have a core of wires, but the wrapping consists of one or more flat metal layers. These outer layers provide much stronger structural integrity and, as a result, flat wrap cables are able to transmit higher tensile and compressive loads even though these flat wrap cables are much less flexible.

Section 3.7

1. The concept of a load acting along a straight line presents the effect on a cable that is supporting a load all along its length. This causes the cable to flex and take the shape of a parabola.

2. The concept of a load acting along a straight line is most clearly demonstrated by the top cable of a suspension bridge. The top cable runs at an angle from the top of the vertical support in the middle of the span of the bridge. It gradually lowers until it reaches the base supports at the level of the bridge surface itself at one of the ends of the bridge. All along this cable, and at consistent vertical spacing, are vertical cables. These vertical cables transmit the load of the bridge uniformly to the cable through the entire length of the cable. Each cable transmits a uniform fraction of the load to the cable. The total load is the weight of the bridge itself. As can be seen in the figure of the free-body diagram, the load (w) is the same at each vertical support cable. This uniform load, with an assumption of a massless cable (the mass of the cable does not enter into the equilibrium equations), creates a parabolic shape in the cable.

3. The formula for the tension in the cable is: $T = T_0 \sqrt{1 + \dfrac{w^2}{T_0^2} x^2}$

   Where:
   w = the load per unit length in newtons per meter (N/m)
   T_0 = the minimum tension on the cable in newtons (N)
   x = the horizontal distance from the left end in meters (m)

   This presents a formula that shows that the tension (T) is equal to the minimum tension T_0 when $x = 0$. Both the load per unit length and the minimum tension on the cable are constants. Therefore as the distance x increases the tension (T) increases.

4. The shape of the cable is: $y = \dfrac{1}{2} \dfrac{w}{T_0} x^2$.

   Where:
   y = the vertical position of the cable from the ground in meters (m)
   w = the load per unit length in newtons per meter (N/m)
   T_0 = the minimum tension on the cable in newtons (N)
   x = the horizontal distance from the left end in meters (m)

   The shape that the cable attains as a result of the uniform load is a parabola.

Section 3.8

1. When talking about a load along a straight line, this refers to a cable experiencing a uniformly distributed load – horizontally uniform distribution. The load is suspended from the cable all along its length. When talking about loads along cables, this refers to the deadweight load of the cable itself with no external forces or loads included in the analysis.

2. When examining just the deadweight load of the cable itself, the shape of the cable is similar to a parabola and called a catenary, but requires a different formula. The word catenary comes from a term used to describe the shape of a chain or rope hanging between two supports.

3. The formula for the tension in the cable is: $T = T_0 \sqrt{1 + \dfrac{w^2}{T_0^2} s^2}$

   This presents a formula that show the tension (T) is equal to the minimum tension T_0 when $s = 0$. Both the load per unit length and the minimum tension are constant. Therefore, as the length s increase the tension also increases by a proportional amount.

4. The shape of the cable is: $y = \dfrac{T_0}{w}\left(\cosh\left(\dfrac{w}{T_0}x\right) - 1 \right)$

Where:
y = the vertical position of the cable from the ground in meters (m)
w = the load per unit length in newtons per meter (N/m)
T_0 = the minimum tension on the cable in newtons (N)
x = the horizontal distance from the left end in meters (m)
cosh = the hyperbolic cosine

When the distance x is zero the vertical displacement y is also zero. The function cosh (hyperbolic cosine function) equals 1 when taken for the value of zero (0); i.e. $\cosh(0) = 1$. The load per unit length and the minimum tension are both constant. Therefore as the distance x increases the value of cosh rapidly increases resulting in a proportional increase in the y-displacement.

Section 3.9

1. A cable with discrete loads is typically a cable that is mounted at the two ends with point loads hanging at random locations along the cable.

2. Three assumptions are applied with cables with discrete loads:
    • The deadweight of the cable is small enough to be ignored.
    • The cable is ultimately flexible and therefore cannot transmit a moment
    • The axial stiffness of the cable is infinite which means the cable will not elongate due to excessive loads

3. Mechanical equilibrium is a condition of an object where the sum of the forces acting on the object is zero and the sum of the moments acting on the object is zero. When considering two-dimensional equilibrium we have mechanical equilibrium: $\sum F_x = 0$, $\sum F_y = 0$ and $\sum M_z = 0$.

4. This requires the equations for mechanical equilibrium of the forces: $\sum F_x = 0$, $\sum F_y = 0$.

    This also requires two equations for mechanical equilibrium of the moments:

    The sum of the moments about one end point: $\sum M_z = 0$.

    The sum of the moments about the random point on the cable: $\sum M_z = 0$.

1. If a liquid is placed in a container the liquid spreads to take the shape of the entire container. This assumes the container is not moving and has no holes in the walls. Under the influence of gravity the liquid molecules will fill practically any shaped container. The liquid will flow in every direction with the end result being that the top surface of the volume will be flat. This will happen with any container of any shape even if the bottom surface of the container is irregular.

2. Using the formulas $P = F/A$ and $F = mg$ we have:
   $P = (500 \text{ kg})(9.8 \text{ m/s}^2)/10 \text{ m}^2 = 4{,}900 \text{ N}/10 \text{ m}^2 = 490 \text{ N/m}^2 = 490 \text{ Pa}$

3. Using the formula $P = F/A$ we have:
   $P = (10{,}000 \text{ kg})/5 \text{ m}^2 = 2{,}000 \text{ N/m}^2$

4. The viscosity of a liquid is a measure of the resistance to flow. Honey has a high viscosity at room temperature and a low viscosity at higher temperature. For a liquid to flow, the molecules must be able to slide past one another.

   In general, the viscosity of a liquid increases with:

   • Stronger intermolecular (between molecules) forces of attraction
   • A greater ability to form intermolecular (between molecules) bonds, especially involving several bonding sites per molecule.
   • Increasing size and surface area of molecules which also increases intermolecular (between molecules) forces.
   • Longer molecules, because they are more likely to get tangled up with one another which makes it harder it is for them to flow.

   The viscosity of the fluid will impact the speed with which it flows and hence the speed with which a defined liquid volume over an area will exhibit uniform distribution of the load. This also impacts how a fluid flows onto or off of a surface. A high viscosity fluid will cause greater resistance to movement and hence a force of drag or friction that is higher than a low viscosity fluid.

5. To be considered an ideal gas a gas needs to pass three tests:

   • The space the individual gas molecules (molecular space) occupy must be insignificant when compared with the space in between the gas molecules (intermolecular space). In other words, the gas molecules can have no significant volume.
   • The intermolecular forces between the gas molecules are generally negligible and therefore can be disregarded in calculations.
   • The gas molecules are usually moving so rapidly that their interactions are not affected by the intermolecular forces.

   Ideal gases do not exist in the natural world, but in most cases, real gases act very much like ideal gases. At extremely high pressures and extremely low temperatures the behavior of gases begins to diverge from that of an ideal gas but this is rarely experienced. Normal pressure is considered 1 atmosphere, the pressure on the Earth's surface on a normal day. Normal temperatures are the standard ambient temperature of about 25°C (298 K).

Section 3.11

1. A sail on a sailboat is pressed by air acting as a fluid. Water towers are built at an elevation so that the water fluid pressure forces the water to flow.

2. The standard units for pressure in SI Units are pascals (Pa) which equals newtons/meter² (N/m²). The U.S. Customary units for pressure are "pounds per square inch" or psi.

3. Pressure = force/area; $P = \dfrac{F}{A}$

   Where:
   P = pressure in pascals (Pa) or newtons/square meter (N/m²)
   F = force in newtons (N)
   A = area in square meters (m²)

4. Pressure = force/area; $P = \dfrac{F}{A} = \dfrac{40 \text{ N}}{10 \text{ m}^2} = 4 \text{ N/m}^2$

5. For a circle the center of pressure is located at the radial center. The resolved load acting at that point is the total load of 150 kN.

Section 3.12

1. In a static (non-moving) fluid, the pressure exerted depends on how far below the surface of the fluid the pressure is measured, the density of the fluid and the gravitational acceleration. The pressure in a static fluid results from the weight of the fluid and does not depend upon the shape of the container, the total mass of the fluid, or surface area of the fluid.

2. $P = \rho g h = \left(1,000 \text{ kg/m}^3\right)\left(9.8 \text{ m/s}^2\right)(0.6 \text{ m}) = 5,880 \dfrac{\text{kg} \cdot \text{m/s}^2}{\text{m}^2} = 5,880 \text{ N/m}^2$

   This equals 5,880 Pa

3. Pascal's Principle states *"A change in the pressure of an enclosed incompressible fluid is conveyed undiminished to every part of the fluid and to the surfaces of its container."*

   The Pascal principle shows that if the pressure is transmitted through the fluid, the force that is applied only depends on the surface area of the fluid to which it is applied. The additional pressure is not dependent upon the volume, density or depth. The increase in pressure is uniform throughout the volume of the fluid.

4. The minimum angle is 0° with the ray pointing directly to the right. The angle increases counterclockwise. When the ray is pointing directly up the angle is 90°. Pointing directly to the left this is 180°. When the ray rotates further back to directly right it is 360° with is also 0° when considering this to be a circle.

Section 3.13

1. In our real world the idea of a displacement is a change of position between time zero (t_0) and some later time (t_1). This change in position is a necessary condition for real work to be performed.

2. The basic engineering explanation of work is a force applied to an object that causes that object to move in the direction of the force. If there is no movement in the same direction as the force there is no work. For example, with circular motion the centripetal acceleration acting on an object produces centripetal force perpendicular to the direction of movement and hence there is no work performed.

   Work: $W = Fd$

   Where:
   W = work in joules (J) or newton-meter (Nm)
   F = force in newtons (N)
   d = displacement in meters (m)

   With real work, there is a real force and a real displacement. The force can be measured and the displacement can be measured.

3. Applying the formula Work: $W = Fd$

   $W = Fd = (30\ N)(50\ m) = 150\ Nm$

4. In virtual displacement, we establish a random coordinate point in space. We imagine that the point has infinitesimal changes in position which are arbitrary in direction and unpredictable in time. This is virtual displacement and can literally be just imagined. It is a theoretical tool of analysis for engineering studies.

5. **On a particle:** The necessary and sufficient condition for the equilibrium of a particle is that zero virtual work is done by all the working forces acting on the object during any virtual displacement. This must be consistent with any constraints imposed on the particle – as discussed above.

   On a rigid body: The necessary and sufficient condition for the equilibrium of a rigid body

   is that zero virtual work is done by all the external forces acting on the object during any virtual displacement. This must be consistent with any constraints imposed on the particle – as discussed above.

   No virtual work is done by internal forces, by reactions in smooth constraints or by any forces perpendicular to the direction of motion. Virtual work is done only reactions when friction exists.

Section 3.14

1. Potential energy is a called the energy of position. This is energy that an object has because of its physical position in a field such as a gravitational field, or by virtue of its position in the reflex action of a spring.

2. Near the surface of the Earth gravitational acceleration (g) is a constant at 9.8 m/s² (32 ft/s²). With mass (m) also being a constant under normal conditions the scale of the potential energy depends only on height. Therefore the gravitational potential energy which uses the formula $U = mgh$ is equal to $U = m(9.8 \text{ m/s}^2)h$ and where mass and gravity are constant the potential energy is dependent upon only the height above a location.

3. The elastic potential energy formula: $U = \dfrac{1}{2}kx^2$

   Where:
   U = elastic potential energy in joules (J)
   x = the distance in meters (m)
   k = the spring constant, a relative measure of the elasticity of the spring in newtons/meter (N/m).

   This is the potential energy that a bob on a spring has when it is at the positions with the largest displacement of the spring, either in compression or in extension.

4. For general gravitational potential energy there is a formula that is more widely applicable than the formula $U = mgh$.

   Gravitational Potential Energy: $U_G = -G\dfrac{m_1 m_2}{r}$

   Where:
   U = potential energy in joules (J)
   G = the universal gravitational constant:
   m_1 and m_2 are the masses of two objects in kilograms (kg)
   r = the distance between the two object in meters (m)

   This applies to objects at a great distance from the earth where the force of gravity is no longer constant.

5. This uses the formula $U = mgh$ which is equal to $U = m(9.8 \text{ m/s}^2)h$. Here the force is given which is the mass × gravity. ($F = mg$)

   At a height of 20 m: $U = Fh = 40 \text{ kN}(20 \text{ m}) = 800 \text{ kNm} = 800 \text{ kJ}$
   At a height of 130 m: $U = Fh = 40 \text{ kN}(130 \text{ m}) = 5{,}200 \text{ kNm} = 5{,}200 \text{ kJ}$

1. First, draw the free body diagram.

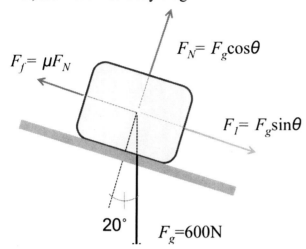

Second, determine the normal for (F_N).

$F_N = F_g\cos20 = (600\ N)(0.94) = 563.8\ N$

From the normal for we can determine both the force of static friction and kinetic friction.

Static Friction: $F_f = \mu F_N = 1.1(563.8\ N) = 620.2\ N$

Kinetic Friction: $F_f = \mu F_N = 0.15(563.8\ N) = 84.57$

2.

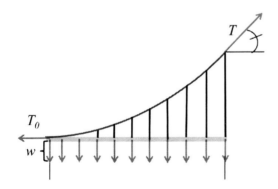

First, the total load is the uniform load (w) measured over the length (x).

$wx = 300\ MN/m(80\ m) = 24.0 \times 10^9\ N$

To determine the value of the tension T, we use the angle is 45°, where the sin45° is 0.707.

Therefore, the tension is: $T = \dfrac{wx}{\sin\theta} = \dfrac{24.0 \times 10^9\ N}{0.707} = 33.9 \times 10^9\ N$

This is the tension of the two cables, so each cable would have ½ of this value. For each cable the tension will be: $1.7 \times 10^9\,\text{N}$.

Again, apply the angle of 45° to find the minimum tension T_0 where $T_0 = T\cos\theta$.

$T_0 = T\cos\theta = 33.9 \times 10^9\,\text{N}\cos45° = 33.9 \times 10^9\,\text{N}\,(0.707) = 24.0 \times 10^9\,\text{N}$

This is the tension in the pair of cables, hence the value of T for each cable is ½ that shown above; i.e. $12.0 \times 10^9\,\text{N}$.

The equation describing the tension is: $T = T_0\left(1 + \dfrac{w^2}{T_0^2}x^2\right)$

Adding the values from above:
The equation describing the tension in both cables is:

$$T = 24.0 \times 10^9\,\text{N}\sqrt{1 + \frac{\left(300 \times 10^6\ \text{N/m}\right)^2}{\left(24.0 \times 10^9\ \text{N}\right)^2}x^2}$$

This can be simplified to: $T = 24.0 \times 10^9\,\text{N}\sqrt{1 + \left(1.56 \times 10^{-4}/\text{m}^2\right)x^2}$

This can be approximated for a single cable as: $T = 12.0 \times 10^9\,\text{N}\sqrt{1 + \dfrac{\left(150 \times 10^6\ \text{N/m}\right)^2}{\left(12.0 \times 10^9\ \text{N}\right)^2}x^2}$

To determine the shape that the cable takes:

The shape of the cables is: $y = \dfrac{1}{2}\dfrac{w}{T_0}x^2$

Adding the values from above: $y = \dfrac{1}{2}\left(\dfrac{300 \times 10^6\ \text{N/m}}{24.0 \times 10^9\ \text{N}}\right)x^2 = \dfrac{1}{2}\left(1.25 \times 10^{-2}/\text{m}\right)x^2$

The shape of the cables is: $y = \left(6.25 \times 10^{-3}/\text{m}\right)x^2$

To determine the shape of the individual cables will require this value to also be divided in half.

The shape of the cable is: $y = \left(3.13 \times 10^{-3}/\text{m}\right)x^2$